Beeching, Wilfred A.

Century of the typewriter

Century of the Typewriter

Century of the Typewriter

WILFRED A. BEECHING
Director, British Typewriter Museum

ST. MARTIN'S PRESS: NEW YORK

For information, write:
St Martin's Press, Inc. 175 Fifth Ave, New York, N.Y. 10010

Printed in Great Britain

Library of Congress Catalog Card Number: 73-87390

First published in the United States of America in 1974

41

This book is dedicated to my darling wife
JEANE
my faithful companion
in life and in business for 34 years –
in appreciation of her encouragement
and patience. Also to my two sons
IAN and MICHAEL
who made this book possible.

25.00

Preface

The typewriter has been with us for a century. Someone had to invent it, as it is so essential to commercial life as we know it today, and played a leading part in the emancipation of women.

For many years I have owned the largest private collection of old typewriters in the world, exhibited in The British Typewriter Museum, 137 Stewart Road, Bournemouth, England. In the course of my study of these machines, I found information difficult to obtain and sometimes contradictory.

During idle hours some years ago whilst on a cruise in the Atlantic, I decided a book was needed on the subject and worked out the general outline it should take. Strangely enough, the typewriter really belongs somewhere in the middle of the Atlantic, as all the designers and producers were either Americans or Europeans.

Most historical writing on the subject is in French, German or Italian, and I soon became overwhelmed with information in four languages. After much sifting and sorting, I was able to produce this book, a comprehensive history of typewriters which are most likely to be seen either in commerce or in various collections. It is a readable story of the typewriter from its beginning up to 1973. It can be read as a story from cover to cover, or used for reference purposes.

I welcome any constructive comments. My hope is that the considerable quantity of data it contains will make this book a source of value and inspiration for the next century.

Bournemouth WILFRED A. BEECHING

'I hear, I balance, I assess, but judge I do not'
– *Ayesha*

Personal Acknowledgments

I express sincere thanks to the following for their kind help and assistance over many years:

Adler, Michael H.: Australian journalist and author, now living in Italy, with whom a personal and helpful exchange of information has been established, and who so kindly read the proofs.

Batchelor, John: author of *Tank* and several other books, for his help and encouragement.

Bliven, Bruce, Jr.: for permission to reproduce information from his privately circulated book *The Wonderful Writing Machine,* which was commissioned by The Royal Typewriter Company of U.S.A.

Bradshaw, George: of The Imperial Typewriter Company Limited, Leicester, for his kind personal assistance based upon his vast experience and knowledge gained over the last thirty-five years.

Brooks, F. M.: formerly Managing Director, Olympia Business Machines U.K., for his kind help and assistance.

Buckeridge, H. W. D. (Buck): a veteran of World War I, for his detailed and intimate personal knowledge of the history of typewriters from 1920 onwards.

Church, W. E.: of the Science Museum, South Kensington, who was instrumental in introducing me to many useful contacts and whose encouragement and help with the basic grouping of typewriters has been particularly invaluable.

Hollis, W. H. (Will): who, for fifty years, was Librarian of The Imperial Typewriter Company, for all his help in compiling my Chapter 1.

Markus, Erich M.: of Office and Electronic Machines Limited, London, to whom I am indebted for a vast amount of detailed knowledge gained in a lifetime of service to the typewriter industry and also for his guidance and advice.

Reed, Frank B.: of Business Equipment, Manchester, for the gift of the book *List of Manufacturing Data of German and Foreign Typewriters*, and for his kind co-operation.

Saunders, Derek: of Facit Office Equipment Limited, Rochester, U.K., for his help.

Sawyer, G. S.: of Remington Rand Division, London, for his co-operation.

I wish to express sincere thanks also to:

Eric Bennett, former Headmaster of Bournemouth School, Hampshire, who helped so extensively with translations and correspondence in many different languages.

My editor and friend, James Negus, for his patience and co-operation.

Winklers Verlag for the loan of their book *Handbook of Typewriting* by Lang-Kruger from their archives, with permission to reproduce information and many invaluable illustrations.

Longs Limited, for the helpful loan of their books.

A. Ransmayer and A. Rodrian, Berlin, for copies of books on typestyles.

Not least are my grateful thanks and sincere appreciation expressed to three long-suffering secretaries! They are:

Mrs. J. Wright
Mrs. J. A. Andrews
Mrs. Louise F. Raymond

W. A. B.

Contents

	page
Preface	vii
Personal Acknowledgments	ix
CHAPTER 1 THE HISTORY OF TYPEWRITERS	1
The first attempts to communicate	1
Early attempts at producing a typewriter	3
The first machines to write	4
First attempts in Europe	7
The dawn begins to break!	10
Mitterhofer's story	18
Pastor Malling Hansen's 'writing ball'	21
John Pratt	23
Franz Xavier Wagner	26
Yost	27
Sholes's contemporaries	27
The first commercially produced typewriter	28
Christopher Latham Sholes	28
The world of 1873	32
Typewriters and their social context	34
The emancipation of women	34
The typewriter in professional life	35
The typewriter during war	37
CHAPTER 2 THE TYPEWRITER TODAY	39
The struggle for the correct keyboard	39
Keyboards for all the world	43
Type and type styles	78
Inking ribbons and carbon ribbons and their various attachments	83
Annual production of typewriters	86
The age and efficiency of a typewriter	87
Typewriters in use today	88
Possible origins of numerals	90
Some odd and interesting facts about typewriters	91
CHAPTER 3 HISTORY OF TYPEWRITER MANUFACTURERS PART I	92
Adler	92
Brother	98
'Erika' – 'Ideal'	100
The 'Facit' Organization	104
Godrej 'All India'	108
Hammond Typewriter and the VariTyper	108
Hermes	113
CHAPTER 4 HISTORY OF TYPEWRITER MANUFACTURERS PART II	122
IBM	122
Imperial	127
Lucznik	131

The Siemag Messa 136
Nippo Portable Typewriter 137
Nippon 138
Olivetti 140
Optima 148

CHAPTER 5 HISTORY OF TYPEWRITER MANUFACTURERS PART III 149
Olympia 149
Remington 152
Royal 157
S.C.M.: The Smith-Corona Story 161
Triumph 169
Zeta and other Typewriters manufactured by the ZbrojovkaWorks in Czechoslovakia 173

CHAPTER 6 ELECTRIC AND SPECIAL-PURPOSE TYPEWRITERS 176
Electric Typewriters 176
Writing Machines for the Blind 183
Noiseless Typewriters 183
Cipher Machines and Secret Writing Machines 185
Pneumatic Typewriters 186
Music-Writing Typewriters 186
Shorthand Writing Machines 187
Toy Typewriters 187
Rebuilt Typewriters 188
Chinese Typewriters 190

CHAPTER 7 MACHINES NOT LOST – BUT GONE BEFORE 191
'Woodstock' (R.C. Allen) Typewriters 192
The Barlock 195
The Blickensderfer Typewriter 199
The Burroughs Typewriter 200
The Continental 201
Everest 202
The Mercedes 203
The Oliver 205
The Smith Premier 208
The Stoewer 209
The Torpedo 210
The Underwood 214
The Yost Typewriter 219

CHAPTER 8 A COMPENDIUM OF TYPEWRITER HISTORY 220
Conclusion 266
Bibliography 268
Acknowledgment of Sources 269
Index 273

CHAPTER 1

The History of Typewriters

The following quotation is attributed to Robert Graves:

'A veteran typewriter of which you have grown fond seems to reciprocate your feelings, and even encourage the flow of thought. Though at first a lifeless assemblage of parts, it eventually comes alive.'

With his thought we wholeheartedly agree. No one who studies typewriters and their fascinating story could ever look upon them as 'lifeless assemblages'.

THE FIRST ATTEMPTS TO COMMUNICATE

It would be impossible to write the history of the typewriter from its actual inception, for there was no true beginning. In order to view things in perspective we have to look at the original efforts of man to communicate.

Throughout the ages, men have had the urge to convey their thoughts, ideas, and hopes, and to record them for those who come after. To this end they have scratched, drawn, painted, chiselled, written, engraved, printed, and finally typed their way through history.

We can begin our story in the Stone Age, when men left their record on the rocks in the form of drawings or paintings. For millions of years before man appeared on this planet, wasps made paper from wood pulp which they used in the construction of their nests. The Egyptians were the first to use it as a writing surface in the form of 'papyrus' from which our English word 'paper' is derived.

Papyrus was made from a rush-like plant which flourished on the marshes of the River Nile. The stems were split, soaked in water, opened out, laid crossways on each other, and dried under pressure. Strips of almost any length could be built-up in this way and the finished

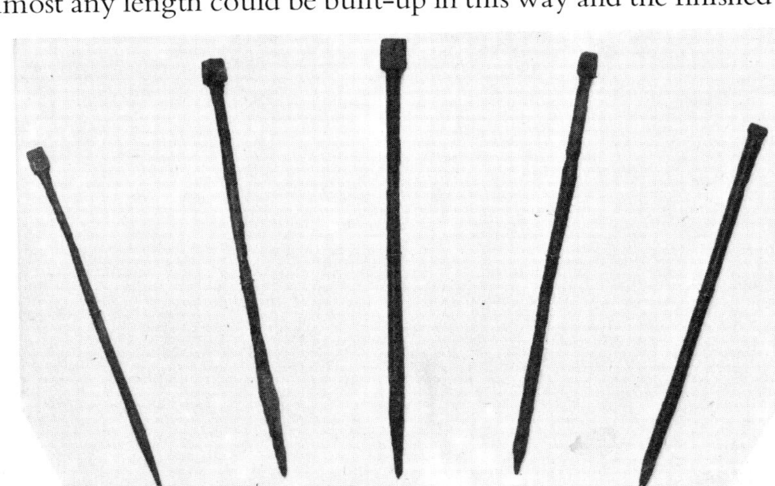

1.1 Roman styli in iron

papyrus was written upon with a sharpened reed, or painted with a brush, using ink made from sepia, charcoal, ochre, and other substances mixed with gum. Many of these documents have been perfectly preserved in the dry climate of Egypt.

During the following 2,000 years, alphabets and methods of recording them were developed by most of the civilized nations. In South-West Asia, for instance, there was a widespread method of writing called 'cuneiform' (Latin *cuneus*—a wedge; from the shape of the impressions). In one form or another this style of writing was used by Assyrians, Babylonians, Chaldeans, Sumerians, Akkadians, Hittites, and Ninevites—civilizations which rose and fell in Old Testament times.

It is believed that the Phoenicians about 1,000 B.C. were the first people to use a complete alphabetic system which was the ancestor of our own. Their numerals in particular were very much like ours. Some students maintain that even these systems were derived from still earlier Minoan forms. However, it is certain that by 400 B.C. the use of the alphabet in one form or another was almost universal.

After the 4th century A.D., the Romans and other European nations used vellum for their more important writings. Vellum was originally a calf-skin and was inscribed on both sides. After use, the inks could be washed or scraped off and the precious material could then be used again.

During the 14th century A.D., parchment came into use as a writing surface. This was also animal skin, although not necessarily that of the calf. Sheep and goat skin could also be used. At first it was used in rolls, later in sheet and book form, written upon with quill pens.

It is not quite clear who first had the idea of using a goose-quill for writing, but it was certainly a great step forward. Originally the whole quill, complete with feather was used, but later the quill was cut into small lengths, sharpened, slit at the end, and fitted on to a stick. Note that our word 'pen' (Latin *penna*—feather) should really be confined to what we call the 'nib', the shaft being the pen holder. Originally too, the 'pen-knife' was a small pocket-knife or similar instrument carried for the sole purpose of trimming the quill to a writing point. To give greater durability the quill sections were sometimes gilded—hence it was an easy step to the manufacture of metal 'nibs'.

We might just mention here that our universal ball-point pen dates back to 1888 while, in one form or another, the fountain pen is very much older.

Printing has probably had a greater impact upon human progress than any other single achievement since the discovery of fire. Printing of designs from carved wooden blocks came to Europe from China in the 6th century A.D., but it was not until the 15th century that the idea of making duplicate copies by inking the raised surfaces of the blocks, was adopted in the West.

Type moulds appeared in 1450 A.D. A Gothic type Bible was produced by Gutenberg of

1.2 Fragment of papyrus from Egyptian ***Book of the Dead***

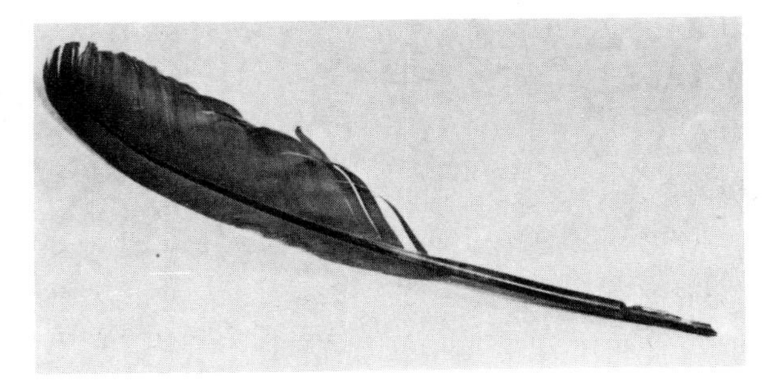

1.3 A quill pen

1.4 Quill pen and pen-knife

Mainz (Germany) in 1455, thus anticipating Caxton by some twenty years. Caxton did not set up his first press in the precincts of Westminster Abbey until 1476.

The typewriter was the logical development of the printing press and about 1750 the stage was all set for a device that would remove the limitations of the written word, and stand midway between the arts of printing and writing.

EARLY ATTEMPTS AT PRODUCING A TYPEWRITER

Over 6,000 years of writing history have now been covered!

As far back as 1647, William Petty, a brilliant and gifted man, socially well connected, a close friend of Samuel Pepys and John Evelyn, and also a Secret Service agent in Ireland, was granted a patent by Charles I, having invented a machine which:

> '. . . might be learnt in an hour's time, and of great advantage to lawyers, scriveners, merchants, scholars, registrars, clerks, etcetera; it saving the labour of examination, discovering or preventing falsification, and performing the business of writing—as with ease and speed—so with privacy.'

This could have been a primitive typewriter, but research at the British Museum shows clearly that it was not a typewriter but some kind of jointed or flexible framework for handwriting with two pens at once. Such a framework was in fact used by the American President Thomas Jefferson a few years later.

Petty's product is only interesting because it was an early device intended to lessen the labour of writing.

James Young was a solicitor, a mechanically-minded man with a very shrewd eye for business. He made an application to the Scottish Privy Council stating that:

> '. . . he had been at great paines and expense in bringing to perfection ane engine (*sic*) for writing, whereby five copies may be done at the same time, which it is thought may prove not un-useful to the nation.'

He was granted a 19 years' exclusive privilege to make and sell his invention to the public.

In the early years of the eighteenth century, the idea of a machine to remove the limitations of the pen was present in the minds of many people.

It is reputed that in 1711 James Ranson made the first typewriter which, we are told, 'amazed all beholders'. *The Yorkshire Post* of 2 May 1952 states:

> 'This machine was in existence up to 1907 when it was shown at an exhibition in London by Mr. Norman Dare. It has (or had) small keys like those on spinets and harpsichords, but each key was painted with a letter in a green, red and blue circle. Long steel rods surmounted by brass squares, engraved with letters, were attached to a steel frame. When the keys were pressed down they struck an inked ribbon. Movement was by means of powerful springs. A long bar held the paper. . . .'

No drawings appear to exist of this machine but one model was made and used for demonstrations. It had an inked ribbon—surely the earliest recorded description of such a device, the next being in connection with Alexander Bain's patent of 1841.

In 1714, the last year of her reign, Queen Anne granted a patent to Henry Mill. It was Patent No. 385 of 7 January 1714 and Henry Mill affirmed that he had:

'. . . by his great study, paines and expense, lately invented and brought to perfection an artificial machine or method for the impressing or transcribing of letters, singly or progressively one after the other, as in writing, whereby all writings whatsoever may be engrossed on paper or parchment so neat and exact as not to be distinguished from print . . . the impression being deeper and more lasting than any other writing, and not to be erased or counterfeited without manifest discovery.'

We know a great deal about Henry Mill, but very little about his actual invention. He was born in Sussex and died, unmarried, in London in 1771. 'His capacity was excellent in all the branches of mathematicks and other liberal sciences. . . .' He was a waterworks engineer at the New River Company, and he designed and erected waterworks in many other parts of the country, including Northampton where he received the Freedom of the Borough in 1722 in recognition of his services. However, no model or drawing of his machine survives.

It would appear, from notes found among Mill's papers after his death in 1771, that he had in mind some method of making raised letters for the use of the blind. This would explain the use of the term 'deeper impressions' and 'not to be erased'. Perhaps it just involved the embossing of a page of print by means of a bookbinder's type under heavy pressure.

THE FIRST MACHINES TO WRITE

In 1830 William Austin Burt produced a typewriter that actually wrote one letter after the other.

He was then nearly 40 years old, and had gained considerable experience over a wide field.

He was born on a farm and, at the age of 12, he quickly showed an aptitude for mathematics and mechanical matters at school. He had an absorbing interest in astronomy and navigation, and wanted a sea-going career. His mother persuaded him against this and he turned his attention to surveying.

At 14 he constructed a sextant with which he worked out the exact latitude of his father's house. Then, at 18, he acquired an old secondhand broken down surveying compass. He repaired it, and used it to help neighbouring farmers to determine the precise boundary of their lands. After a brief spell of military service, he married Phoebe Cole, whose father he worked for at one time.

He practised his trade of millwright, journeyed through the States, and finally settled in 1824 in a remote part of Michigan, living in a 'do-it-yourself' log cabin, working as a blacksmith, making his own tools, and continuing his interest in surveying. He became a Justice of the Peace, Postmaster and County Surveyor and, eventually, became Deputy Surveyor of the United States and a world-recognized authority on navigation.

It was here, in the backwoods of Michigan, that he constructed his 'Typographer', as he called it, made largely of wood, using only tools which he himself had made.

We must give him full credit for originality, as although several European inventors were working along the same lines, Burt could hardly have had access to any reports of their efforts.

The photographs in Figures 1.6*a* and *b* are not of the actual machine for it was destroyed in a disastrous fire at the U.S. Patent Office in 1836. Fortunately, the original parchment, drawings, and specifications, escaped the fire and from them a faithful replica was built in 1892 by one of Burt's descendants. This was placed in the Smithsonian Museum and another replica was

1.5 William Austin Burt (1792–1858), inventor of Burt's Typographer

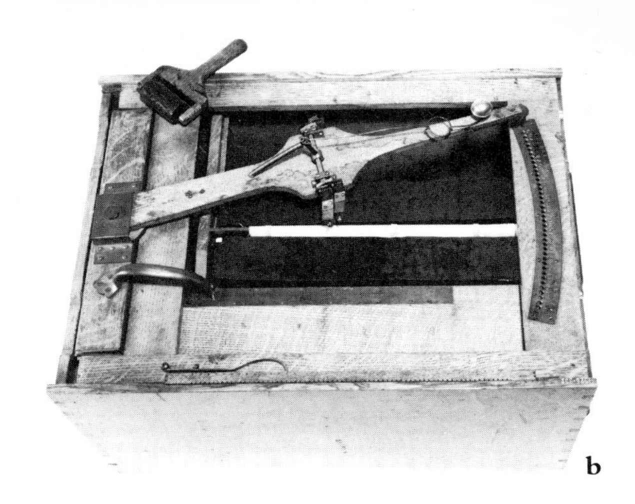

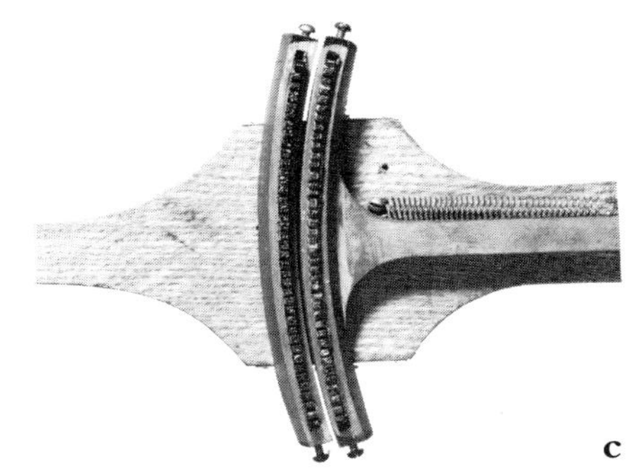

1.6a and b Burt's Typographer of 1829 (reproduction); c shows type on swinging sector

made for the Science Museum at South Kensington, London, and is the one illustrated.

The Typographer, patented in 1829, featured type mounted on a semi-circular frame which was brought to the printing point by turning a wheel mounted on the face of the instrument. The 'clock face' on the front of the box was actually the line-spacing lever and the spacing could be varied by means of interchangeable racks. The type was obtained from a friend, John Sheldon, editor of the *Michigan Gazette*. As soon as the problem of type was encountered, John Sheldon offered to supply printer's type from the newspaper office. This he did and Burt fitted the type to the sector and the 'Typograph' became a writing machine. The type was mounted on a semi-circular frame which was brought to the printing point by turning a wheel mounted on the face of the instrument. Printing was achieved by depressing a small lever (on the left of the face in Figure 1.6*a*) which brought the type snugly against the surface of the paper. Ink was supplied by two inking pads located on either side of the printing point.

As soon as he saw that the machine really worked, Sheldon took a hand in the proceedings. Fired with enthusiasm, he borrowed it and 'typed' the following letter to Senator van Buren:

'Hon. M. van Buren,

Detroit, N.T.
May 25, 1829.

Sir,

This is a specimen of the printing done by me on Mr. Burt's Typographer. You will observe some inaccuracies in the situation of the letters;

these are owing to the imperfections of the machine, it having been made in the woods of Michigan, where no proper tools could be obtained by the inventor, who, in the construction of it, merely wished to test the principles of it, therefore taking little pains in making it.

I am satisfied from my knowledge of the printing business, as well as from the operation of the rough machine, with which I am now printing, that the Typographer will be ranked with the most novel, useful and pleasing inventions of the age.

Very respectfully,

Your obedient servant,

John P. Sheldon.'

Burt himself typed the following on the back:

> 'I, William A. Burt, being duly sworn, depose and say that I am the inventor of the machine, called by me the TYPOGRAPHER, and intended for use in families, offices and stores, and further that such invention and any parts thereof, have not, to the best of my knowledge and belief, been known or used in the United States or any foreign country.'

Burt obtained his patent, signed by President Jackson and Senator van Buren on 23 July 1829. He then built another on which he wrote a letter to his wife, Phebe.

A copy of this letter typed on the improved machine is illustrated here (*see* Figure 1.8) and the machine itself was rather an odd instrument. It looked rather like a 'pin table' mounted on four carved legs.

Burt then seemed to leave his 'Typographer' and returned to his original love, inventing a solar compass which would remove the errors caused by the unreliability of the ordinary magnetic variety. He was successful in this, and the new compass was patented, becoming standard equipment for the U.S. Government surveys.

It is significant that the world was not ready at that time for a typewriter. Even had it

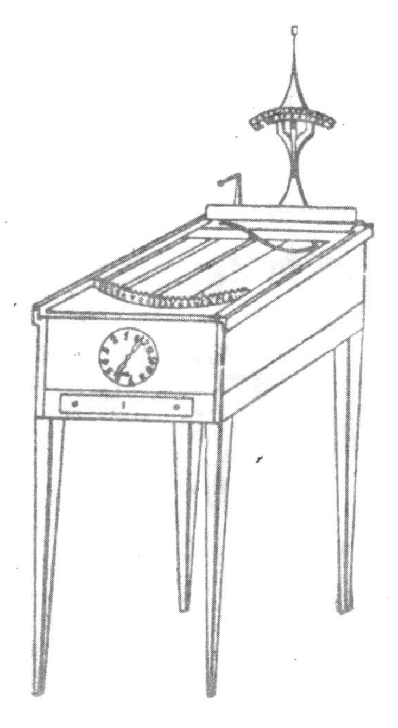

1.7 **Burt machine of 1829**

New-York March, 13, 1830

Dear Companion,

I have but jest got my second machine into opperation and this is the first specimen I send you except a few lines I printed to regulate the machine, I am in good health but am in fear these lines will not find you so andthe children from the malencholley account your letter gave me of sickness and deaths in our neighbourhood, I had rested contented to what I should if it had been summer season adout the health of my family, as it is jenerlly healthy during the winter months; but their has ben an unusual quentity of sickness heare this winter, and it has ben very cold in Urope as well as in America, a strong indication of the change of seasonth that I have so often mentioned.— Mr Sheldon arrived here four days ago he went imediately on to Washington and took my moddle for the Pattent Office, he will returne here next week at which time I shall put my machine on sale and shall sell out the pattent as soon as I can and return home, at ancy rate I seall returne home as soone as the Lake nevigation is open if life and health is spared me. I have got along but slow since I have been here for the want of cash to hire such help as I wanted; I have been as prudent as I could, have taken my board with a family from Cyuga who keep a bording house they are verry good christian people and are kind to me. I pay three Dollars a week for my board.—You must excuse mistakes, the above is printed among a croud of people asking me maney questions about the machine. Tell the boys that I have some presents for them. If I had aney news to communicate I would print more but as I have none I must close hopeing these lines will find you well. I wish you to write as soon you receive this, do not make aney excuses I shall like see it in aney shape

William A. Burt.

Phebe Burt

1.8 **Burt's letter to his wife, 1830**

been, the principle adopted by Burt, that of the 'linear index' was far too slow to compete with the keyboard-operated typebar machines which were later to appear.

However, Burt certainly left his footprints in the sands of time, although not in connection with typewriters.

His solar compass was standard equipment for more than 75 years; indeed, in a modified form it is still in use today.

His third invention—the 'equatorial sextant' was patented in 1856 and was received with enthusiasm by navigators generally. At the time of these inventions, they must have seemed to Burt of far greater significance than the 'Typographer', which is probably the reason why he did not follow it up.

He died in Detroit in 1858 as a result of a seizure when instructing a class of sea-going captains in the use of the equatorial sextant.

FIRST ATTEMPTS IN EUROPE

While Burt was experimenting in America, there were many inventors in Europe interested in the idea of a mechanical writing machine. Even before Burt's time, in the year 1753 the German von Knauss invented a 'writing machine', of which he constructed no less than four different models between that year and 1760. These, we are told, 'achieved a limited success and aroused both interest and emulation'.

Note particularly the word 'emulation'. It is not without significance, for we have records of several other attempts in Vienna during the next few years.

Von Knauss was a professor of Physics, and occupied the important position of Director of the Physical and Mathematical Laboratory in the Austrian capital.

The next recorded invention was that of Count Leopold von Neipperg of Vienna. A full description of this machine together with excellent contemporary drawings exists in the Imperial and Royal Library at Vienna, but it was a typewriter that was never really produced.

An illustration in the *Export Review* of September 1924 leaves no doubt about its true structure. It was a rather large, elaborate, and ornamental framework, raised to hand height on Jacobean legs, for writing on two or three sheets of paper simultaneously with mechanically linked pens. This was produced in 1762.

Pierre Jaquet-Droz of Neuchâtel in Switzerland was the next inventor although he was really a copyist or improver. He adapted the machines of von Knauss and von Neipperg, embodying their best parts in a machine of his own.

The first authentic record of any machine which would be called a 'typewriter' was produced by a Hungarian, Wolfgang Kempelen.

He was a superintendent of the salt mines, an inventor, a poet, a scientist, an artist, an architect, and a linguist who spoke nine languages. He invented a chess-playing automaton which beat the best players, a 'speaking machine' and a typewriter.

A description of the typewriter is quoted from *Office Appliances* of October 1912:

> 'Kempelen had successfully experimented with the box shape, with hidden key levers. Under the patronage of a favourite of the Empress Maria Theresa, he continued his researches, and at length brought out a machine which wrote the German alphabet in large, clear type—about the size of modern large primer—several specimens of which are still exhibited in a Vienna museum. This machine also came into active use with the beginning of the nineteenth century. It spaced accurately, punctuated amply, and typed clear, firm impressions of each letter. In fact, apart from its still clumsy construction the typewriter had now become a really practical business convenience, and among the limited number of commercial houses which could afford its purchase, was so regarded and used.'

Kempelen, like Leonardo da Vinci, must have been far ahead of his time. His life and inventions are of absorbing interest. His automatic chess-player achieved international fame. Napoleon (himself a good chess-player) tried his skill against it when visiting Schönbrunn. It beat him and he became so furious that he had to be restrained from attacking his automatic adversary.

Kempelen also improved the Savery steam-engine, constructed a special sick-bed for the Empress, made a writing machine for the blind, and invented an elaborate system for teaching the deaf and dumb. He must have been a most intriguing character!

In 1808 a young Italian named Pellegrino Turri of Castelnuovo, enters the story. He was friend and protégé of the Countess Carolina Fantoni da Fivizzono, who was blind from her youth, but nevertheless very active and energetic. She carried on a copious correspondence and felt a pressing need for something which might assist her with it.

Turri commenced work and built her a writing machine which was the first to print in anything like the impression of a modern typewriter.

Count Emilio Budan of Venice wrote in his brochure *Precursors of the Modern Typewriter* (1912):

> 'Kempelen's typed "copy" was simply that of the printing-press of those days—thick, black characters, clear to read but in no way distinctive.
>
> Turri's machine, on the contrary, built in 1808, wrote entirely in capital letters, thin, distinct and easily read to this day in the Italian Archives of State.'

Apart from this meagre information there seems no trace of any further detailed particulars of Turri's contribution.

In 1823, Pietro Conti of Cilavegna, Italy, claimed that he had invented a machine that 'wrote fast and clear enough for anyone, even those with poor sight'. It was taken to France, patented, demonstrated before the Académie Française, tested by them, approved, and eventually purchased for 600 francs.

In November 1934 a tablet was unveiled in Italy dedicated to 'Pietro Conti. the inventor of the first typewriter'. Conti had called his machine 'The Tachigraph' (Greek *tachos*—speed, and *graphein*—to write).

In 1830 an Italian artist called Galli devised a 'Mechanical Potenografo'. This, it seems, was a machine with keys corresponding to the letters of the alphabet arranged in two concentric circles. The types struck downwards towards the paper which was coiled on a drum. This revolved between the letters which were ordinary roman characters.

Galli brought his machine to England where it was received with great favour, as is shown by the following article in *The Times* newspaper of 27 June 1831:

> 'A very ingenious piece of mechanism has lately been invented by a young Italian gentleman of the name of Galli, now in this country. One of its objects is to enable us to write faster than any system of shorthand hitherto known, or than any orator can speak. But this is not all. Many copies of a discourse, legibly written, may be taken at the same time. . . . It is played upon by the fingers, like a musical instrument, and the manuscript is rolled off a cylinder during the course of the writing. By employing it, a book may be copied while the reader is perusing it, and as fast as it can be read. The judge on the bench may, by its means, take down the depositions of witnesses while his mind is intent upon the hearing of the evidence. By a little habit even the blind may be made to use an instrument which will enable them to copy faster than any shorthand writer. This ingenious machine has many other advantages, which, if realised according to its inventor's expectations, will produce great changes in our present system of written or telegraphic communication.'

'Many copies . . . may be taken at the same time' could mean nothing less than some form of manifolding. It could also imply the use of carbon paper. Unfortunately, no further information can be traced of either Galli or his machine, but he certainly seems to have been on the right lines, even though he did not succeed in selling it.

Next we have Xavier Progin of Marseilles who filed patent No. 3748 in France for his 'Machine Ktyptographique' on 6 September 1833, but this is important enough to be dealt with later.

In 1837, the year of Queen Victoria's accession to the throne, the typewriter was virtually unknown, but by the end of her reign, it had developed to such an extent that a gold-plated machine was delivered to Buckingham Palace for her own personal use.

In France a machine was made by Gustave Bidet of Paris. He was a watchmaker and, surprisingly enough, he was the first watchmaker who seems to have been interested in 'Writing Machines'. Watches and clocks had been constructed efficiently for many centuries and it is quite curious that watchmakers had not been involved in this kind of development before. Like Progin, Bidet also patented his machine which actually worked.

No such machine appears to exist today and, in 1909, George Carl Mares, in his book *The History of the Typewriter* states:

> '. . . it possessed at least four of the essentials of a good writing machine. It consisted of two posts, carrying at the upper portion a wheel, on the periphery of which were the types. These types, as they revolved on the wheel, took a supply of ink from two small rollers, one on either side. Below this wheel was a cylinder for carrying the paper. . . .'
>
> 'Evidently the cylinder revolved between each letter and made a side movement between the lines. The wheel had to be revolved by hand until the required letter was over the printing-point, when pressure upon a lever which was borne on the axle of the type-wheel, brought the letter into contact with the paper, and thus secured the impression. A writing machine, certainly—but a very slow one.'

This description would almost exactly fit the Columbia type-wheel machine produced in 1886 which achieved wide popularity over a considerable period of time.

Giuseppe Ravizza of Novara in Italy started experimenting with typewriters in 1832–3, but he is dealt with more fully later.

In France in 1838, Monsieur A. Dujardin built two devices, one of which was for recording speeches and the other for printing. The first of these had twenty-six keys resembling those of a piano, arranged in alphabetical order, with a letter at one end and type at the other.

However, this was completely impracticable as the paper was limited to four inches in width and there were no numerals or punctuation marks.

In Rouen in 1839 Louis Jeŕôme Perrot produced a machine that printed designs on cloth. From this he progressed to letters on paper. In August 1839 he took out a patent for 'Machines and devices for typographic, lithographic and tachygraphic impressions, with a special view to use on paper'. Budan states:

> '. . . (the typewriter) was perhaps the first in Europe to offer anything like an embryonic keyboard. A vertical cylinder, endowed with an intermittent movement of a rotatory nature, was its principal feature. Two metal wheels bore the alphabetical characters. Crude of course, but in general design a tolerable facsimile of those in use today. The cylinder was so constructed that paper could be readily slipped on and off.'

The main interest in Perrot's activities is that he is stated to have 'disposed of a number of his machines among Rouen commercial houses'. A typewriter salesman on the road in 1839 staggers the imagination!

Baillet de Sondalo, a Parisian, invented a machine in 1841 which he called 'The Universal Compositor'. It worked by means of an indicator, something like a clock-hand which was moved over a circle of characters by a pedal operated by the user's foot. When the indicator came over the required letter a lever was pulled and this depressed an engraved type-sleeve or cylinder into contact with the paper.

In a model in 1841 the foot-pedal was eliminated, the indicator being operated by the left hand while the right hand brought the type-cylinder into contact with the paper.

De Sondalo claimed that his 'Universal Compositor' would '. . . replace all typographical material, and even the pen, and would also render unnecessary the study of shorthand'. Most optimistic!

Baron Drais von Sauerbronn is said to have invented a machine with sixteen square keys. An early article on typewriter history tells us that it was invented by a 'bicycle manufacturer'. This 'bicycle' was actually what we call the 'hobby-horse', ridden by the dandies of the period. It had no pedals, chain or other mechanisms but was propelled by the rider's feet pushing alternately on the ground, free-wheeling downhill with legs outspread.

This vehicle was named the 'Draisine' after its inventor, and a very good example of it is preserved in the Science Museum at South Kensington.

We know all this about the 'bicycle' that was invented but nothing at all about the 'typewriter with sixteen square keys'!

THE DAWN BEGINS TO BREAK!

The story continues with Progin in France and Thurber in America, separated by thousands of miles in space and ten years in time. It is extremely unlikely that either of them had any knowledge at all of the other's invention. Had it been otherwise, the world might have had a practical typewriter long before it actually did.

Firstly, Progin of Marseilles, France, a practical printer, had ideas about mechanical writing many years in advance of his time. His machine is of great interest, because it incorporated the first recorded use of individual linked typebars operated by leverage and converging to a common printing-centre.

From the illustration here (*see* Figure 1.9) it can clearly be seen that the Ktyptograph belonged to the 'down-strike' class, but it did not have a carriage. In fact, the writing was done on a flat surface. Such a machine has its advantages. It is one thing to write on a single sheet of

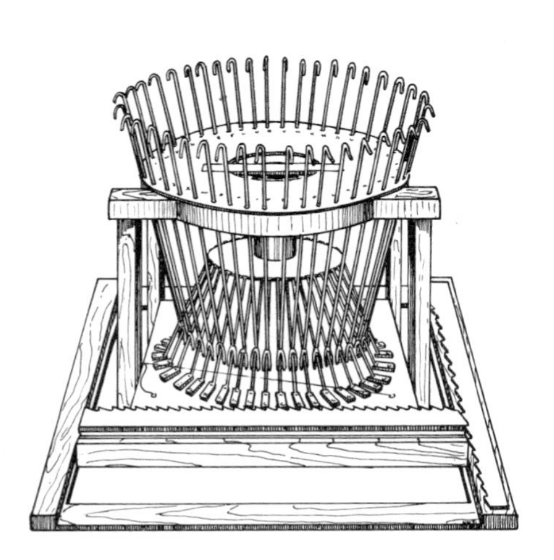

1.9 Xavier Progin's Machine Ktyptographique of 1833; he invented the principle of the typebar

1.10 Thurber machine of 1845

NORWICH 3. FEBRUARY 1846

GENT.

WE HAVE, AT LENGTH COMPLE-
TED ONE OF THURBERS MECHANICAL CHIROGRAPHERS. ALTHOUGH YOU
WILL NOTICE IMPERFECTIONS IN THE FORMATION OF THE LETTERS
IN THIS COMMUNICATION, YET THERE IS NOT A SINGLE DEFECT
WHICH DOES NOT ADMIT OF AN EASY AND PERFECT REMEDY. I
AM PERFECTLY ~~SATISFIED WITH IT BECAUSE I DID NOT LOOK FOR~~ I
PERFECTION IN THIS FIRST MACHINE. THE DIFFICULTY IN THIS MA-
CHINE IS THAT THE CAMS ARE NOT LARGE ENOUGH. THIS, OF
COURSE, CAN BE AVOIDED. I THINK MR. KELLAR TOLD WHEN I
LAST SAW HIM THAT IF I WOULD WRITE TO HIM INFORMING HIM
WHEN I SHOULD BE IN WASHINGTON HE MIGHT BE ABLE TO
MAKE SOME SUGGESTIONS ABOUT A HOME DURING MY STAY IN
WASHINGTON. I SHALL WISH TO EXHIBIT THE MACHINE. TO SUCH
GENTLEMEN AS MIGHT ~~TAKE INTEREST~~ IN A THING OF THIS
KIND. I DO NOT WISH TO ~~MAKE~~ A PUBLIC SHOW OF MY-
SELF OR MY MACHINE. I WANT ~~TO SHOW~~ IT ~~TO MEN~~ WHO CAN
APPRECIATE AND UNDERSTAND MACHINERY. MR. ROCKWELL. OUR REP-
RESENTATIVE IN CONGRESS VOLUNTEERED TO GET ~~ME~~ A ROOM &
I HAVE WRITTEN TO ~~HIM~~ ON THE SUBJECT. STILL I THOUGHT IN
CONSEQUENCE OF YOUR MORE THOROUGH ACQUAINTANCE IN THE
CITY THAT YOU MIGHT BE ABLE TO MAKE SOME SUGGESTIONS
WHICH MIGHT BE BENEFICIAL TO ME IN EXHIBITING HE MACHINE.
I WANT A ROOM LARGE ENOUGH TO RECEIVE SUCH COMPANY AS
MAY WISH TO SEE THE MACHINE. I WANT A ROOM WHERE I
CAN SAFELY LEAVE IT WHEN I AM ABSENT AND WHERE NO
ONE WOULD BE LIABLE TO GO IN AND INJURE IT. EXCUSE THE
LIBERTY I HAVE TAKEN, AND BELEVE ME

YOURS, TRULY. CHARLES THURBER.

MESSRS. ~~KELLER & GREENOUGH~~
PATENT ATTORNIES.
WASHINGTON. D. C.

1.11 Facsimile of first letter written on Thurber's machine dated 1846

paper that can be rolled around a cylinder, but quite another to write in a bound book without removing pages or on a sheet of stout card without bending it. Progin's machine could do this and in this respect it foreshadowed the well-known Elliott-Fisher typewriters which came into being some sixty years later to start a long and successful history.

There was no keyboard, just a circle of stout wires with their upper ends turned over to enable them to be pressed down by the operator without injuring the finger tips. The lower end of each wire was forked and slid along the typebar depressing it and bringing the type block down to a common central printing point. It is curious that a printer such as Progin did not take the obvious step of substituting lettered buttons or keytops. This would have enabled the operator to depress the right letter at sight. As it was he had to look at a separate guide-card, hunt for the corresponding hook, press it, and hope for the best!

The Ktyptograph was very slow in operation although Progin himself in his patent No. 3748 of 6 September 1833 in France, stated that his machine 'would write almost as fast as a pen'.

What Progin in fact devised was the all-important function of individual typebar leverage —a principle underlying the action of nearly every make of typewriter on the market today. He constructed a typewriter for typing music and in some small way this actually outlined the basic principle of the shift key.

The second person involved was a man by the name of Charles Thurber. He was an American of some standing who became obsessed with the idea of mechanical writing. Working in Worcester, Massachusetts, he built several models, the first being in 1843 and the last in 1845. All these belonged to the circular index class, although, as the complete set of types had to be moved in order to obtain one impression, they could also fall into the 'block' category.

His 1843 machine worked on the same principle as the old machines found on railway platforms which produced a strip of thin metal on which we could emboss our names or to be more up to date, a primitive Dymo machine which is in common use in offices today.

Thurber's 1843 model used a single impression lever which passed through a lettered slot but in its final form (*see* Figure 1.10), the machine was provided with a separate plunger for each character. One of these machines is preserved in the Smithsonian Institution in Washington.

The most striking advance of Thurber's 'Chirographer' was its carriage and it was the first machine with a cylindrical platen. It had a rack and pinion escapement, but like all machines working on the various 'index' principles, it was slow in operation. However, the 'index' principle, whether circular, linear or square, slow as it was, was copied and re-copied during the succeeding seventy years.

Many of the machines made on this principle around the middle of the century were intended to emboss characters for blind readers, a purpose for which the principle was entirely adequate. It also appeared in toy form but for business purposes it was much too slow.

By 1846 Thurber had grown tired of typewriters. In that year he made a massive machine which actually wrote the letters with a pencil! It was a cumbersome machine, the size and shape of an old-fashioned hand-loom or stocking-frame. It wrote upon a vertical sheet of paper in full view of the operator. There was a horizontal shaft in front carrying an appalling collection of cams, two for each character, and we understand that it 'was operated by power'. All this, just to propel a pencil over a piece of paper! But, the machine actually worked. Only one of these machines was ever built and we can quite understand why.

1.12 Hughes's Typograph for the blind of 1850

1.13 Pierre Foucault's Printing Keyframe or *Clavier Imprimeur* of 1851

In 1851 two men quite independently produced a successful typewriter to enable the blind to communicate with each other and to record their thoughts in a form which could be read by sighted people. Although they were working completely independently they both produced their typewriting machines in the same year and they were shown at the Great Exhibition of 1851, where each was awarded a prize medal, and each was subsequently adopted and used by Blind Institutions.

At this time a Mr. Hughes was Director of the Henshaw Institute for the Blind at Old Trafford, Manchester. He obviously realized that the blind could read matter embossed by the method of raised dots invented by Louis Braille in 1834, but they had no method by which they could express or record their own ideas. With this end in view he devised and built a machine which he called the 'Typograph'. The machine itself was a marvel of simplicity, but it worked, and it worked well. His first model produced embossed characters which the blind could read by touch, but a later model was adapted for printing ordinary letters on paper.

By 1851 machines produced by Hughes were in daily use in schools for the blind. A *Minute Book* for that year contains some neatly-typed and beautifully-spaced entries, mostly inventories of supplies, and it also records that when Queen Victoria visited the Institute she witnessed a demonstration by Mary Pearson, one of the blind pupils, and was filled with astonishment and admiration at the facility with which the girl typed the phrase 'Her Most Gracious Majesty'.

The 'Typograph' is also mentioned in the Institute's *Annual Report* for 1851, wherein the

Board placed on record their appreciation of 'The Director's ingenious invention for enabling blind persons to communicate with others, whether similarly afflicted or not'.

In 1843, Pierre Foucault, a pupil of the Blind Institute in Paris, was working along much the same lines. He constructed machines which produced embossed letters. There were ten sliding pistons, each carrying six types. Foucault called his machine the 'Raphigraphe'.

Foucault later abandoned this and built another machine with a separate key for each character arranged in two double rows. A blank key was also provided for spacing between words. These banks of keys were 'radial' or fan-shaped so that all the types converged to the same printing-point on the paper where they made their impression through a sheet of carbon paper.

Like Progin's machine the paper was held in a flat frame underneath the machine, but here the resemblance ceases, for whereas Progin's paper was fixed while the whole machine moved across it, Foucault's apparatus reversed the process. The paper-carrying frame moved horizontally and automatically for letter-spacing and vertically by hand for line-spacing.

After many trials and errors between 1843 and 1849, each an improvement on the last, he finally made a machine in 1850 which was shown to, and praised by the Paris 'Board of Encouragement' which awarded him a gold medal.

In 1851 his last perfected model was brought to London and shown at the Great Exhibition in Hyde Park, where the Board of Examination unanimously awarded him a gold medal for his invention.

A great deal of space was devoted to this machine in the National Press in France and Italy, and after the Exhibition many 'Clavier Imprimeurs' were sold for the equivalent of £20.00 each, being used by Blind Institutions for a number of years.

It is interesting to note that while both Hughes and Foucault started off with the simple idea of producing embossed letters for the blind to read by touch, each of them, after a short interval, realized the extended scope which might be given to their machines, and produced models using ordinary printing type, thus enabling the typewritten work to be read by sighted individuals.

The alleviation of the lot of the unfortunate sightless appears to have been the principal thought behind the design of most writing machines from the time of Turri onwards. It appears strange that none of these early inventors seem to have realized the enormous potential value of their inventions in the freeing of business, industry, and literature from the tyranny of the pen. Even when the first efforts were made in this direction the promoters came up against such a blank wall of prejudice that the 'Typewriter' was introduced only with the greatest difficulty into the dingy gas-lit offices of the early eighties. With regard to personal correspondence, a certain amount of this prejudice still exists. Many elderly and aristocratic individuals maintain that to type a letter to one's friends is one of those things which is just 'not done'.

How strange it is that back in 1830 blacksmith Burt knew nothing of this social stigma. An ordinary affectionate individual, he typed a letter to his wife Phebe addressing her as 'Dear Companion', but then he was a distinctly uninhibited person.

The moral eludes us . . .

During the period between 1850 and 1860, there seems to have been a halt in the development of typewriters in America. In England research and experiments went on continuously, and many successful working models were being built by various inventors in other countries. We shall probably never know how much we owe to European typewriter inventors who, after doing the initial research, seemed content to leave the subsequent development, manufacture, and marketing to enterprising American industrialists with their superior facilities.

Very little appears to have been attempted by anyone until Christopher Latham Sholes started experimenting in 1867. Even Scholes's efforts are believed to have been prompted by

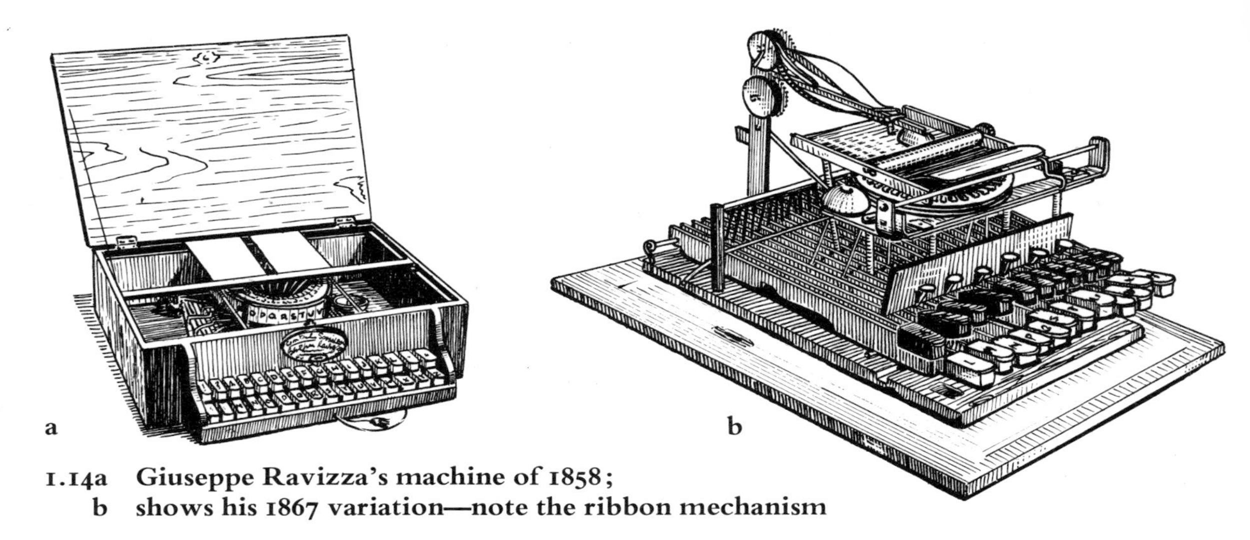

1.14a Giuseppe Ravizza's machine of 1858;
b shows his 1867 variation—note the ribbon mechanism

an article in *The Scientific American* which described a machine designed by American-born William Pratt which was perfected and manufactured in London.

In Novara in Italy, Giuseppe Ravizza invented and experimented with seventeen different models of 'Writing Machines'. He was a solicitor in his native town, a mechanically minded man and versatile. In 1832, at the age of 19, he commenced working on the first of a long series of experimental model typewriters.

In 1837 he produced a machine but it received very little attention. Not discouraged though, he persevered and by 1855, his work was sufficiently advanced to enable him to take out a patent for what he described as a 'macchina da scrivere a tasti', that is a 'Writing Machine with keys'. This was shown at the Industrial Exhibition at Novara. Unfortunately, this model was later destroyed by his grandchildren who used to amuse themselves playing with the 'toys' he had constructed with so much care and patience. But it was minutely described in the Patent, so Ravizza was able to build another which he entered at the much more important 1858 Exhibition in Turin, and it was awarded a Silver Medal for merit.

He called his first machine the 'Cembalo Scrivano' or 'Writing Harpsichord' on account of its superficial resemblance to that musical instrument.

All his machines were of the up-strike variety arranged in a full circle, and the writing was not visible. In 1860 his personal diary records that he had plans for the construction of a true 'visible writer', although this did not appear in his patents until 1883, unfortunately too late. Its prototype had been exhibited at the Milan Exhibition of 1881.

His 1867 model had many refinements. At the end of a line a little door opened and a visible signal popped out which said 'The line is finished', at the same time a bell sounded. This model had partly visible writing.

There is not the slightest doubt that Ravizza was one to make practical use of the travelling inked ribbon. Previous inventors had used sheets of carbon paper. The ribbon was able to move along and across, thus using every part of the inked surface, and it was charged with aniline dye dissolved in glycerine to defer drying out. He did not go into production with his machines as he was not satisfied with any of the ones he had made. He sold a few machines which he had built himself; all of them were different with improvements each time.

In 1883 he applied for a comprehensive patent, only to find that nearly all his original ideas had been incorporated in machines on the other side of the Atlantic, particularly on the Model 2 Remington to which his final model bore an almost uncanny resemblance. By this time the American machines had been in commercial production for almost ten years. Ravizza was frustrated and naturally disappointed, but he carried on with his experimental work until his death in 1885 at the age of 74, a year after the appearance of his seventeenth and final model.

In 1931 a monumental stone was unveiled at Novara commemorating his work, and a special medal was struck to mark the occasion. An illustrated brochure was also published, claiming that it was Ravizza's work which led ultimately to the construction of the Sholes and Glidden typewriter, but there is no means of substantiating the truth or otherwise of this statement.

The most unlikely people seem to have tried their hand at typewriter making. So far we have had Burt—a blacksmith; von Knauss—a professor of physics; von Neipperg—an Austrian Count; Kempelen—who was everything; Turri—who does not appear to have been anything in particular; Galli—a young teacher; Progin—a printer; Bidet—a watchmaker, Drais who made hobby-horses; Thurber—a mechanic; Hughes—Director of the Blind Institution; Foucault—a blind pupil, and Ravizza—a lawyer.

We shall now deal with Wheatstone, an electrician, Hood—another blacksmith, Guillemot—an instrument maker, Flamm—another printer, and Dr. Francis—a physician.

In the year of the Great Exhibition in 1851, when Foucault and Hughes were making machines for the use of the blind, a British inventor named Charles Wheatstone, Professor of Experimental Physics at King's College, London, was embarking on a brilliant career devising means for the transmission of electric signals over wires, inventing the dial telegraph, the 'Wheatstone Bridge' familiar to all electricians, many automatic transmitting and receiving instruments, and the type-printing telegraph.

It is in connection with the type-printing telegraph that his activities come into our particular sphere. Wheatstone was not interested in the typewriter as a universal writing instrument, but only as a convenient method of recording incoming telegraphic messages on tape.

His first machine of 1851, had the conventional 'piano keyboard' of the period. The arrangement of the letters was alphabetical and the type was mounted on the flexible teeth of a comb, radiating from a common centre.

He brought out his second model in 1856, and this had a shift key, and printed both upper and lower characters. This model also typed on a flat surface. It is not generally known that all modern typewriters are made with the type curved to correspond with the convexity of the platen. Were this not the case then the impression from each type would tend to be heavy at the centre and light at the top and bottom of the letters. Before this method was adopted, the earliest typewriter builders tried to overcome the difficulty by:

1. using a flat surface;
2. using smaller type;
3. increasing the diameter of the platen.

It must be explained that very complex manufacturing operations are necessary in order to achieve this result and this was not within the powers of the early makers.

Charles Wheatstone was knighted in 1868 for his services to science. He died on 19 October 1875, having invented the rheostat, the magneto-exploder, the concertina, the harmonium, and his telegraphic apparatus. His physical researches included investigations into the transmission of sound vibrations through solid bodies, the spinning mirror for measuring the velocity of an electric discharge, and the physiology of vision which later led to the invention of the stereoscope. A truly versatile and remarkable man!

In 1857, Peter Hood produced a typewriter. His father was a blacksmith in Kirriemuir in the county of Angus, Scotland. Peter followed in his father's footsteps, but ill-health forced him to give up this strenuous work and he became a watch and clockmaker. This trade he carried out in the attic of a little cottage where he lived in semi-seclusion with his two unmarried sisters. Here, in addition to watches and clocks he built many original and highly ingenious mechanisms.

1.15 Charles Wheatstone (1802–75)

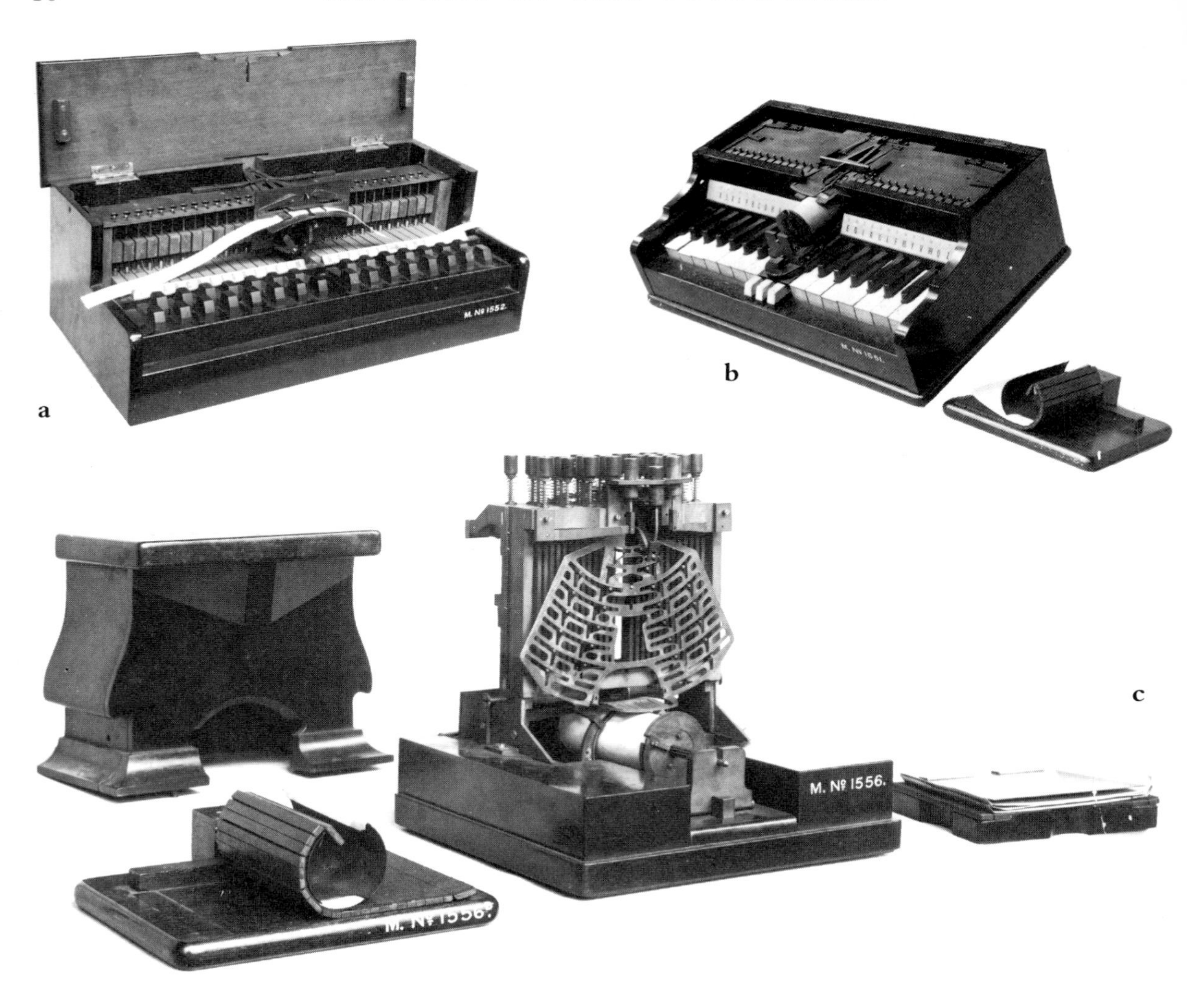

1.16a Wheatstone Typewriter of 1851
b another model
c his Typewriter of 1856

One of his patrons was a certain Mr. Arrol, a relative of Sir William Arrol. This Mr. Arrol was blind, and Hood was commissioned to build a typewriter for him which Hood did with some success. He was without any previous knowledge of typewriters, and he started from scratch. He recognized the four basic functions which must be incorporated in any writing machine:

1. to bring the type to a common printing point;
2. to move the paper step by step;
3. to ink the type; and
4. to render the writing visible.

He built two machines in 1857, one of which is known to have been sent to America and the other is in the Science Museum at South Kensington, but unfortunately, in a damaged condition.

Like all 'index' typewriters it was slow in operation but it was a true typewriter, resembling the Columbia type-wheel machine of which many examples still exist. Peter Hood died in 1873, the year recognized as the commencement of typewriter production.

Adolphe Charles Guillemot of Paris produced a typewriter in 1859. He was a designer of

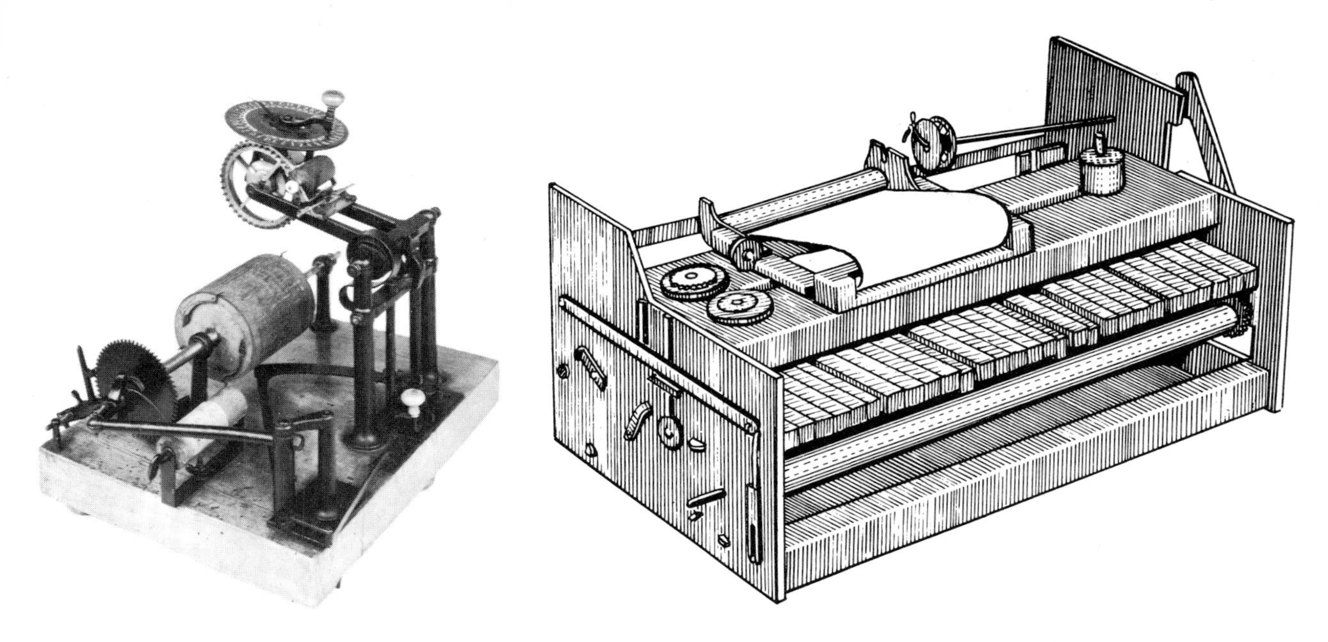

1.17 Hood Typewriter of 1857

1.18 Guillemot's machine of 1859

precision instruments and had better qualifications than most of the other inventors of his time to design a typewriter. His French patent states:

> 'A machine for corresponding or quoting, etcetera, and made so that a person not able to write, or suffering from infirmity can use it by simple contact with the notes which print a certain number of letters and thus record his thoughts.'

He suggested that his machine could be used by the blind. Through all the years, the inability of all these inventors to see the commercial possibilities of their invention, is staggering. Their attempts to cater for the blind or infirm are highly commendable.

Whether the Guillemot machine is still in existence is a matter for conjecture. It was displayed as recently as 1924 in an Exhibition where it was shown to the Chambre Syndicale de la Mécanographie in France. However, a drawing does exist.

Poor Guillemot was unsuccessful in raising capital in France to develop his machine, so he brought it to England in 1861, meeting with no greater success.

Monsieur Flamm was a printer. His machine worked broadly on the principle of the Linotype machines and anticipated by some twenty years the invention of this machine by Ottmar Mergenthaler. Flamm had no idea of applying his invention to printing. What he was after was a machine that would produce typewritten matter a line at a time. In this he was successful. His idea was excellent, but the world was not ready for such a sophisticated invention and the principle has no parallel amongst today's machines.

Dr. Francis was a physician who lived in New York. In 1857, during an illness he contracted, he found himself with a lot of spare time and, in his enforced idleness, he became attracted to the idea of developing a 'Writing Machine'. He was no mechanic and, therefore, looked around for advice for a practical solution to his problem.

He took his rough drawings to a carpenter and explained his requirements, but the carpenter was no 'Mitterhofer' and soon became discouraged. Dr. Francis then tried a musical instrument maker who specialized in the making of harps. This person expressed great interest but was not prepared to undertake this work, believing it to be beyond his capabilities.

Finally, Dr. Francis employed William Beaumont, a skilled mechanical engineer. Beaumont

1.19 Francis's machine of 1857

had the right approach. He studied the theory carefully, then prepared working drawings from the Doctor's rough plans and eventually constructed the first machine.

It proved to have a close resemblance to Ravizza's second machine. Clearly there had been no collaboration and it seems to have been an instance of two inventors working entirely separately in different parts of the world and achieving the same result.

It will be clear at once from the only photograph known to exist, that Dr. Francis (like so many other early typewriter designers) had fallen under the fatal fascination of the piano keyboard. This is obviously unsuitable for a typewriter where no chords are ever required, but only individual letters each following the other.

Mitterhofer's Story

1.20
Peter Mitterhofer (1822–93)

Peter Mitterhofer, an Austrian, was born in 1822 at Partschins, near Merano, in the Tyrol. His father was a carpenter, and on leaving school, Peter worked with him for a number of years. He had developed a taste for music, and perhaps with some inherited skill as a handyman, he made a guitar for his own use, and as time went on he made other musical instruments, using almost any tool which happened to be lying about. So musical was he, that he gave concerts in the summer time to the villagers, and his contributions included not only musical items but also ventriloquism.

After serving his apprenticeship with his father, he left home to widen his experience, and his travels took him to Germany, France, and Switzerland. He remained a bachelor for a long time, so that it came as something of a surprise when, at the age of 40, he not only married, but married a girl six years his senior.

His bride was a Miss Marie Steidl, who was not without money, for she had inherited a house and a small farm from her parents who were also carpenters.

It was Peter Mitterhofer's ambition to construct a typewriter, and soon after his marriage he set about achieving his ambition. Undoubtedly the marriage he had contracted gave him a certain measure of financial independence, and although he was never wealthy, he was able, for a few years at least, to concentrate on his work without undue financial worries. Altogether he built four models of typewriters between 1864 and 1868 and received a total of 350 guilders towards the cost of his experiments from the Imperial Court of Austria.

Neither before, nor perhaps since Mitterhofer, has anyone had such a logically sound and far-sighted conception of a typewriter, as had this Tyrolean carpenter, and yet in spite of his inventions he died without achieving either fame or fortune.

In an autobiographical poem Mitterhofer stated that 'a carpenter invented the typewriter in Merano in 1864'. It seems certain therefore that his first model dates from 1864—and is identical with the model in the Vienna Technical Museum. It had no platen, but a flat writing plate, made of wood, with a movable frame, and the outline of the letters was formed by needle points puncturing the paper.

Mitterhofer was not satisfied with his first model and realized its shortcomings. He soon began work on his second model, which is now in Dresden, and is inscribed 'First successful model'.

According to the autobiographical poem already mentioned, Mitterhofer was more encouraged by this second model, which differed from the first in that the paper was now held in a frame, so that the letters being printed could actually be seen. A small holder prevented the sheet of paper from being torn as the keys came up from underneath.

Today there is little doubt that it was his third model of 1866 which he took on a home-made wheelbarrow to Vienna—a distance of some 350 miles. There he intended to show that his machine would be useful to solicitors as well as to those working in the Imperial Administrative Offices.

Shortly before Christmas 1866, Mitterhofer wrote a letter to the Emperor, Franz Joseph I, saying:

> 'The most obedient suppliant is the inventor of a writing apparatus which has a great usefulness and is quite unique in its characteristics.'

He emphasized that his apparatus was a prototype and needed certain technical improvements, but that he was only a carpenter, without the money to buy proper equipment. His request for financial assistance was granted and he received a subsidy of 200 guilders for further developments and improvements.

1.21 Mitterhofer's machines 1864–9

It is not known how Mitterhofer reacted to this grant, but he was impressed by the fact that it came through within a month—unusually swift for a Government Department!

This third model is known as the 'Merano Model' and is in the proud hands of the Director of the Town Museum in Merano. It had a shift key, so that the thirty-six keys gave a total of seventy-two characters. In 1867 Professor Heger of the Vienna Polytechnic was convinced that 'this machine might prove useful if it were further developed'.

In January 1870 a second petition was taken to the Emperor in which Mitterhofer said he had made good use of the 200 guilders and had now finished work on another model. The petition closed with the hope that the Emperor would grant him another small subsidy or would arrange for his machine to be purchased as a collector's piece.

At this time the Emperor was beset with more pressing problems, particularly with the problems of war, and he would have preferred to have been shown an invention for a successful war machine rather than a writing machine.

He therefore showed little interest in Mitterhofer's story, particularly when one of his advisers suggested that the machine did not really live up to the inventor's claims and was hardly capable of being put to practical use. In spite of such remarks, however, Emperor Franz Joseph agreed to a suggestion made by a member of his cabinet to give Mitterhofer 150 guilders for the machine which the Emperor soon donated to the Vienna Polytechnic.

Mitterhofer held the view that his typewriter would be invaluable in Government offices and pointed out its potential use for typing secret documents, and with his petition he enclosed a list of advantages he claimed that typing held over ordinary handwriting. These were, briefly:

1. It was quicker than writing, was uniformly clear, and always legible; it took up less space—thus saving paper.
2. There was no strain on the eyes, or chest (as was inevitable if you were crouching over a piece of paper); hence people suffering from chest complaints or eye strain need have no fear about losing their jobs. Moreover you could type in the dark—and even blind people could learn to type in a few days.
3. Very little physical effort was needed, and the absence of physical strain was particularly important for all those engaged in occupations and professions requiring mental energy, e.g. solicitors, diplomats, writers.
4. The typewriter was easily transportable, took up little space, and was therefore particularly useful in war time when military units were on the move. No pen or ink were needed; it could be used in all kinds of weather. It was very good for secret documents as the lid of the machine could be closed down on the writing, thus preventing unauthorized persons from seeing what had been written. It was also very useful for officials who wanted their documents to be read easily, but were only too well aware of their own illegible handwriting.
5. It could be used by the sick, and those confined to their beds. Even those unfortunate people with only one hand could use it.
6. It was invaluable in business, and for those who for one reason or another found writing difficult. It could also ease the work of compositors in printing works, for the machine produced work of great clarity.

Mitterhofer should not have limited his appeals for help to the Emperor, but should have looked elsewhere. However he may well have felt his life-work was virtually finished when he had completed his fourth model. This is surprising, for there is a letter typed by Mitterhofer on 3 August 1869 on his fourth model in which he says 'My typewriter still writes very badly but I have not yet given up hope, for I know how to improve so many things if God will give me time and health'.

He died in 1893, having made no effort apparently to find anyone to manufacture his machine.

Peter Mitterhofer was years ahead of his time with his inventions, and his ideas of construction were quite unique. His typewriters show arrangements which can be found on many of our present day machines. In fact, no other inventor has had so many of his ideas incorporated in today's machines—for instance, variable line spacing.

If Mitterhofer's models had come into the hands of the right engineers at the right time, it is highly probable that, with only simple modifications and different materials, an earlier start would have been made on the manufacture of typewriters, and the resulting streamlining of offices would have come about much sooner; for Remington in 1873 had not really arrived at the stage of development reached by Mitterhofer ten years earlier.

Yet for years he remained in obscurity. Not until the 1920s, after the discovery of one of his early models, was Mitterhofer's work brought to light. Dr. R. Granighstaedten-Czerva, a lecturer in Vienna, tried to rescue Mitterhofer's reputation, and produced a brochure on him in 1924. Mitterhofer was hailed as an unrecognized genius and a story was circulated that an American mechanic, Glidden, had copied a Mitterhofer typewriter which he had seen in Vienna.

This story seems to have been started by one of the inhabitants of Partschins, where Mitterhofer was born, but was later disproved. However, even in 1924, Mitterhofer's grave was inscribed with the words 'The others who learnt from him were able to harvest the fruits of his talents'.

The story, with suitable embellishments no doubt, enjoyed a wide circulation, so that by 1940 it had even appeared in some encyclopedias; but it was finally killed and repudiated by Glidden's daughter. In 1939 she gave testimony that her father had never been to Vienna, had never even left the United States, and had no knowledge at all of Mitterhofer.

Several others, after all, had tried to make a typewriter before, and it was really just a case of two inventors working along the same lines, completely independently of each other. This was certainly true in the case of Mitterhofer for in his autobiographical poem he said 'I can say with pride that the apparatus is my invention and I, a simple Tyrolean peasant from Partschins have never seen anything that could serve as a basis for my model'.

With the outbreak of World War II, Mitterhofer sank back into obscurity, but in 1961, Dr. Joseph Nagler, Director of the Technical Museum, Vienna, renewed the attempt to discover what had happened to Mitterhofer's machines, especially as he found a machine which bore the inscription 'Typewriter of an Austrian Inventor, 1842?'. In 1969 Dr. Nagler gave this model to a member of his staff, Richard Krcal, who had been working on the restoration of earlier Mitterhofer models. When the work of restoring this unidentified model had been completed, the typing it produced was compared with the typing found on the letter from which we have already quoted, written by Mitterhofer on 3 August 1869 and still preserved in the castle of Spauregg, and Dr. Nagler came to the conclusion that the model, previously thought to have dated from 1842, was in fact Mitterhofer's last machine.

Fanciful stories have surrounded Mitterhofer's life—such as the legend concerning an American visitor in about 1870; the secret cupboards which Mitterhofer was alleged to have had in his house; the peasant children playing with the early models, and the Director of the Vienna Technical Museum rescuing them. Such stories appear to be without foundation, and we prefer to accept as authentic the account published by Basten International to whom we are indebted for much of our information about this inventor.

Pastor Malling Hansen's 'Writing Ball'

Pastor Rasmus Hans Johan Malling Hansen was born on 5 September 1835 in Hunseby in Denmark. When he was 4 years old, his father, who was in the teaching profession, died of typhus. The young boy was very industrious and had a special interest in mathematics. On leaving school he was apprenticed to a painter and later studied theology. He eventually became a teacher in a Deaf and Dumb Institute of which he became Headmaster, from 1862 to 1864.

In 1864 he went to Copenhagen, completed his studies, passed his final examination, and was appointed Head of The Royal Deaf and Dumb Institute in Copenhagen.

In his search for a new alphabet for the deaf and dumb, he noticed that, by 'speaking' with fingers, about twelve syllables could be communicated in a second, compared with only four syllables in ordinary writing. So he hit upon the idea that it should be possible to make a machine to enable the deaf and dumb to communicate with each other as fast as they had done previously in sign language.

a

c

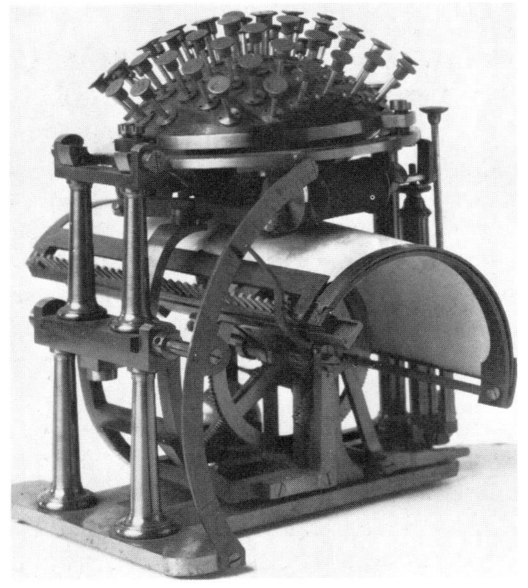

b

1.22a Pastor Malling Hansen's Writing Ball of 1867–68
b another model of the same period
c an electric model of 1870–75

His first experiments were made on a porcelain ball on which he wrote block letters. These experiments lasted several months and in the winter of 1865–6, the Malling Hansen 'Writing Ball' was produced but he did not patent it until 1870. He exhibited the machine and won awards in Copenhagen in 1872, in Vienna in 1873, and an improved model won a Gold Medal at the Paris Exhibition of 1878. He went on trying to perfect the mechanism, and most of his efforts were concerned with improving the movement of the paper through the machine. The general design of the machine was convex to enable all the bars to reach a central point.

The 'Writing Ball' already possessed the most important characteristics of our modern typewriters. It was the first machine of which this could be said, and it was manufactured and used in commercial concerns in Denmark, Germany, Austria, and France, even if the number of machines sold was small.

It had originally fifty-two, and then later, fifty-four keys for signs and capital letters only. It had no type levers, but a vertical or oblique thrust. In 1867 the movement of the carriage was electrically produced. Hansen soon abandoned the electrical movement. The paper was moved by pressing on the keys as in modern machines.

In 1878 a ribbon superseded the carbon paper which was originally used. This ribbon ran from one roll to the next, and could be reversed manually. The chief disadvantage of the machine was the fact that only octavo paper could be used, but this difficulty could have been overcome.

The Nordic Telegraph Company used a number of 'Writing Balls' to reproduce incoming telegrams.

'Writing Balls' are (or were), in the German Museum in Munich, in the Blind Institute in Berlin-Steglitz, in the Museum of the Wanderer-Werke, and also in the Olympia Office Machine Works at Erfurt.

John Pratt

John Pratt was born at Union in the State of Carolina in 1831. While he was still a boy his parents moved to Greenville, Alabama, where he studied law, and they later moved to Centre, Alabama, where he began practising law on his own account.

Amidst the hatred and bloodshed of the American Civil War he began building his 'Pterotype', presumably as a hobby. During the Civil War it was, of course, impossible for a Southerner to take out a patent in Washington. Furthermore, it was not the best background for law practice or typewriter research.

Pratt left his family with his parents and took his machine across the Atlantic. Early in 1864, he took out a provisional British patent, and started building experimental models in Glasgow in conjunction with a scientific instrument maker and a firm of piano manufacturers.

In 1866, he abandoned the 'plunger' principle and procured another British patent, this time for a machine of the 'block' category.

He enlisted the assistance of a mechanical engineer in London, E. B. Burge, and together they made several models using the type-wheel plan.

There is one priceless example of the 'Pterotype' in almost mint condition in the Science Museum at South Kensington and, as far as we know, it is the only one of its kind in existence.

This machine was the forerunner of the later 'Hammond' and 'Crandall' typewriters.

In 1867, Pratt exhibited the 'Pterotype' to the London Society of Arts, and personally lectured on it. The lecture was fully reported and public interest was aroused. Articles appeared in the scientific and technical press, giving rise to considerable correspondence and controversy. Pratt sat down at the 'Pterotype' and wrote a letter to the United States Commissioner of Patents; this letter is still in existence.

During Pratt's stay in England, the bitter struggle in America between North and South came to an end—so too had Pratt's resources, and in 1868, he returned to America. He found that in 1867, on 6 July, the leading technical journal of the United States *Scientific American*, had printed a full description of his machine, together with illustrations of it. They stated that, 'not only would the inventor of a successful Writing Machine confer a benefit on all mankind, but would also incidentally, reap a fortune'. This prophecy proved true in the first part, but as for the fortune—there never has been, and as far as can be seen, there never will be any fortune to be made in the inventing or making of typewriters.

The editorial continued as follows:

1.23 Pratt's Pterotype of 1866

'A machine by which it is assumed that a man may print his thoughts twice as fast as he can write them, and with the advantage of legibility, compactness and neatness of print, has lately been exhibited before the London Society of Arts by the inventor, Mr. Pratt of Alabama. The subject of typewriting is one of the interesting aspects of the near future. Its manifest feasibility and advantage indicate that the laborious and unsatisfactory performance of the pen must, sooner or later, become obsolete for general purposes. Legal copying, and the writing and delivering of sermons and lectures, not to speak of letters and editorials, will undergo a revolution as remarkable as that effected in books by the invention of printing, and the weary process of learning penmanship in schools will be reduced to the acquirement of the writing of one's own signature, and playing on the literary piano above described, or rather on its improved successors.'

This particular paragraph in a paper of 1867, contains the first mention of the word 'typewriting' that we can trace. Commonplace though it is today, it appears to have never before been used. Works such as 'mechanical writing', 'typography', 'print writing', 'embossing', 'tachograph', 'cryptography', etc., had always been mentioned previously. Never 'typewriting'.

On returning to America, Pratt followed up his letter with a personal visit to the United States Patent Office, and was eventually granted U.S. Patent No. 81,000 for his type-wheel principle. Now began a long series of disappointments. His apparent success proved to be 'Dead Sea fruit' for he found that only a few months earlier, Christopher Latham Sholes had patented a machine which, although working on a totally different principle, earned him the honour of being named as the inventor of the world's first commercially-produced typewriter.

He continued with his experimental work and in 1880 he again applied for a patent for a further improved model. This time he was unfortunately forestalled by Lucien Crandall who had secured a patent for his type-sleeve principle a year earlier. But this was not all.

James B. Hammond who had been a newspaper correspondent during the Civil War, had also been working along similar lines in the 1870s. Hammond was the rare combination of a businessman and an inventor, and he took the precaution of protecting his rights although he had not produced a single machine. Hammond offered Pratt a cash payment and a royalty to stay out of the typewriter business. Pratt accepted the offer.

The first 'Hammond' model appeared in 1884, and it embodied several of Pratt's devices plus the revolutionary idea of interchangeable type, effected by means of removable and replaceable shuttles. The present 'VariTyper' is the legitimate descendant of the early 'Hammond' and is the only practical machine on the market today working on the shuttle principle.

The inventions we have mentioned in the preceding pages were only some of the many arrangements and writing machines invented prior to 1873. There were ten more in France, two in the United States of America, and one in Italy before 1850, when Oliver T. Eddy of Baltimore, Maryland, invented a magnificent machine with seventy-eight typebars in thirteen rows, as complicated, and about the same size as a baby-grand piano.

In 1852, John M. Jones of Clyde, New York invented a typewriter that did exceptionally good but slow work. He called it the 'mechanical typographer' and in 1854, R. S. Thomas of Wilmington, North Carolina invented a 'typograph'.

In 1856, John Cooper of Philadelphia patented a big, rugged machine which had a proper paper-feed mechanism and Alfred E. Beach who was the editor of the *Scientific American* patented a 'Writing Machine' which he later billed in his magazine as 'the original typewriter'.

In 1868, when Christopher Latham Sholes and Carlos Glidden with S. W. Soule got their patent accepted, thirteen more typewriters were invented, eight of which were by Americans including F. A. de May of New York, Benjamin Livermore of Hartford, Vermont, George House of Buffalo, New York, Abner Peeler of Webster City, Iowa, and Thomas Hall of Brooklyn, New York.

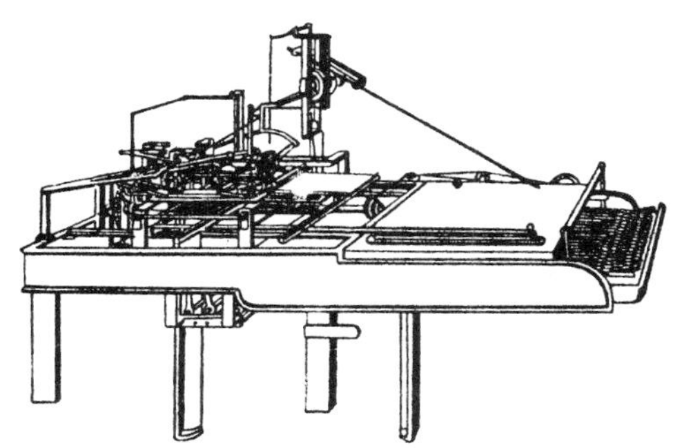

1.24 Eddy's machine of 1850

1.25 John H. Cooper's Radial Strike Machine of 1856; this appears to be the only existing model and is now in the Milwaukee Museum, U.S.A.

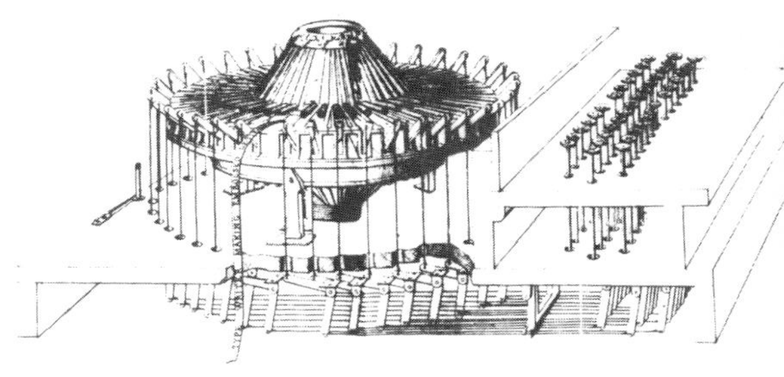

1.26 Beach's Typewriter of 1856

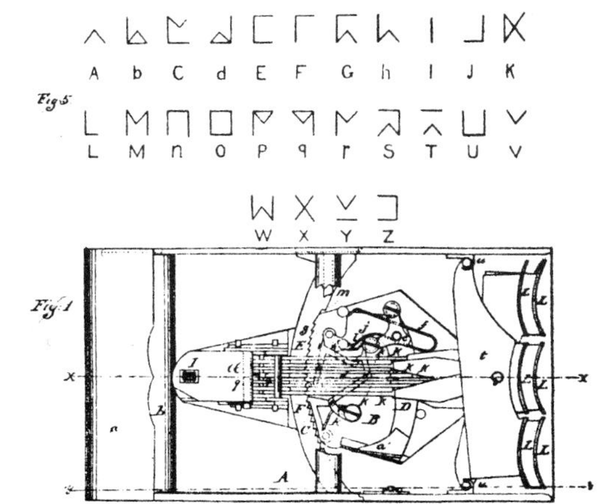

1.27 Livermore's Printing Device of 1863 (patent drawing) showing the character combinations (above)

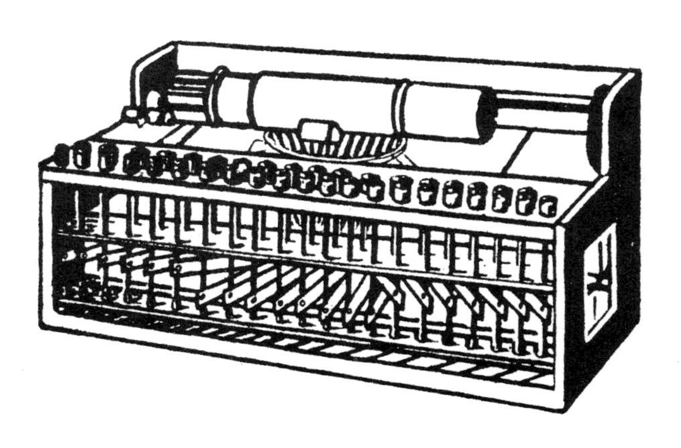

1.28 House's Typewriter of 1865

1.29
Franz Xavier Wagner
(1837–1907)

Franz Xavier Wagner

Franz Xavier Wagner was born on 20 May 1837 at Heimbach, Germany, and was orphaned at the age of 12. This probably made him more self-reliant than most children. He was trained as a mechanic and eventually settled in Stuttgart in 1860, where he not only constructed a sewing-machine but established a business to manufacture them.

In 1864 he left Europe and went to America. During the first years of his stay there, he worked as an ordinary mechanic and in his free time busied himself with all kinds of technical inventions; in fact, he sold patents for the improvement of many tools. About 1870, he invented a hydrometer which won a competition and thus enabled him to sell the rights.

He took a great interest in the construction of typewriters which were then in their infancy. Wagner worked in his own workshop which he himself had designed and equipped. He used no powered machinery, even the lathes being operated by a foot treadle. He had a valuable assistant in his son whom he had taught the basic engineering skills.

Wagner devoted most of his life to inventing and designing typewriters. He was a genius with a fertile and agile mind. He was always trying out new ideas, and was really the outstanding designer of his time and the brains behind most of the leading American typewriters. He did not take up a permanent position in any company, but collaborated with other inventors in the construction of new machines. For instance he collaborated with Densmore to produce in 1891 a machine which bore Densmore's name.

Wagner was also one of those who worked on the construction of the Remington 1873 typewriter as well as on the Caligraph and the Yost and, most important of all, he designed the Underwood.

Wagner had, for many years, been concerned with making the writing completely visible. On most machines which were claimed to be visible, only a part could be seen and most of it was concealed. Only a few machines such as the Daugherty were really visible but they had other defects. Wagner's son invented the linkage between the typebar and the key lever which, with variations, is in general use today. His father related this to the principle of the segment and type guide and thus, between them, they devised the idea of a segment and bars—an idea which has been incorporated in all successful standard and portable typewriters ever since.

Soon Wagner set about making experimental models and trying out his invention. He offered his machine to the Remington Company in 1897, but they turned it down as their own machine with its invisible writing was good and had become very popular; consequently they were not prepared to alter their model.

Wagner wanted to produce something which would completely revolutionize the typewriter and he founded The Wagner Typewriter Manufacturing Company in New York to produce his new model and joined forces with the Underwood Typewriter Company. Once the machine had been thoroughly tested, it was produced in 1898 as the first completely visible writing typewriter with forward-strike.

For about ten years Franz Wagner continued to put his ideas at the disposal of the Underwood Company. With his death in 1907, the typewriter world lost a brilliant engineer.

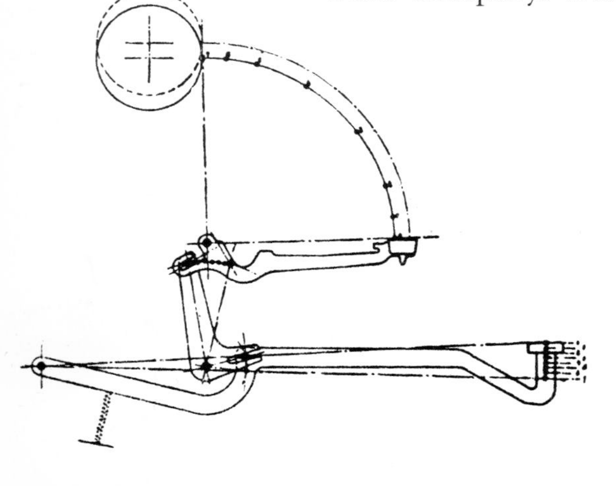

1.30 Wagner's design showing segment and bar

Yost

1.31 George Washington Newton Yost (1831–95)

George Washington Newton Yost was well known for his energy, vision, and enterprise as a salesman, and he worked in the petroleum industry in Pennsylvania before Densmore drew his attention to the typewriter in 1870. When Densmore joined up with Sholes to start negotiating with the Remington Company to produce the Sholes model, he took Yost with him as he was known to be a good organizer and, thanks to Yost, the negotiations reached a satisfactory conclusion.

Yost joined the Remington Company and soon became one of the most valuable members of a small team. He pointed out many defects in the machine and made proposals for its improvement.

Unfortunately he did not always meet with the approval of the Directors, particularly when he proposed to build a machine with a double keyboard instead of a shift lock mechanism. The Directors were not prepared to consider this, particularly as their first models were still in the course of production and it would be very costly to re-equip, so, after working for them for five years, Yost left the Remington Company in 1878. He went on with his ideas, however, and founded a company to manufacture the 'Caligraph', the name of the Company being The Caligraph Patent Company, and he took Densmore and Sholes to work with him.

Yost took over the direction of the manufacture and, at the same time, the direction of the 'American Writing Machine Company' which was to sell the machine, and eventually both of these companies were merged. In 1885 he left the Company.

Yost was not the man to be depressed by one misfortune; he would always set about designing another typewriter. In contrast to the other machines of the time, his next machine, the 'Yost' had, instead of a ribbon, a cushion on which the types were embedded. It also had a double keyboard, and the first model was thoroughly tested for three years before it came on the market in 1889, to be unsurpassed for the next twenty years or so. Yost died in 1895.

Sholes's Contemporaries

Before proceeding to the story of Christopher Latham Sholes, it is only fitting to mention the names of other important people concerned with the development of typewriters. They are not necessarily in chronological order, but were inventors who made contributions to the typewriter as we know it today. Some of these men did, in fact, have their names on machines which were produced for a short period of time, whilst others were purely inventors.

The paths of several of these are repeatedly crossing each other as in the case of Wagner, Yost and Kidder. Some such as Glidden, are connected with only one man (Sholes) or one typewriter, whilst others, such as Blickensderfer, seem to have been relatively independent operators in the field.

Blickensderfer designed and produced the first portable machine and is dealt with in greater detail in Chapter 7 on typewriters no longer with us.

James Denny Daugherty of Kittanning was the man who invented the first forward-strike machine as we know it today, but it lacked certain essential details. The machine was heralded as 'The Daugherty Visible' with the slogan 'we claim everything in sight'. It had four rows of keys and forty keys making eighty characters in all, and it had a shift mechanism. As it was easy to manufacture it could be sold at a modest price.

Originally this machine was produced in 1890 in the Crandall factory under the supervision of E. E. Barney, and later produced by the Daugherty Typewriter Company.

In 1898 Daugherty left the firm. The factory remained in Kittanning but the headquarters moved to Pittsburgh and the machine's name was changed to Pittsburgh; later it was known as the 'Reliance-Premier' the 'Fort Pitt' and finally the 'Schilling'.

Whilst his greatest contribution was the forward-strike machine as we know it today, it was left to Wagner and his son to perfect the idea.

Wellington Parker Kidder was born in 1853 and showed an early interest in engineering. Even at the age of 15, he had patented a device for improving the steam-engine. Later, in 1887 he produced the Franklin typewriter and after five years, the Wellington and the Empire.

In 1896 he entered into an agreement with the Adler Cycle Company which had previously been known as Heinrich Kleyer to produce the Empire typewriter in Germany. This was the beginning of the Adler machine of which, by 1913, 100,000 had been made. Kidder also designed the noiseless action which was incorporated in both the Remington and the Underwood typewriters so it could be said that, although his actual name apparently did not appear on any machine, he was responsible for the basic designs and principles of typewriters which lasted for many years.

James Densmore, of whom we shall hear much more in connection with Sholes, also took part in other typewriting companies, but the Densmore typewriter of 1891 was not, in fact, his own work—it was the invention of his brother Amos Densmore, in conjunction with Sholes and Wagner, but James financed the enterprise.

1.32
Christopher Latham Sholes (1819–90)

THE FIRST COMMERICALLY PRODUCED TYPEWRITER

Christopher Latham Sholes

We will now pass on to the early development of the Sholes-Glidden machine which was the first production typewriter. There were many others responsible for individual developments and the typewriter was not invented by one man alone, although Christopher Latham Sholes received the credit for it.

Christopher Latham Sholes has been described as 'the most unselfish, kind-hearted and companionable man who ever lived'. He was certainly versatile, having been successively a printer, newspaper editor, postmaster, politician, and state senator. He had the light blue, far-away eyes of a visionary, and with his long flowing hair was a familiar figure in the streets of Milwaukee in 1866. He was the fifty-second man to 'invent' the typewriter!

With Carlos Glidden he collaborated in various enterprises, and had recently been working on an agricultural implement, a kind of plough which he called a 'mechanical spader'. Together with a local printer, Mr. Samuel Soule, Sholes had successfully developed a machine for the automatic consecutive numbering of railway tickets, which could also be adapted for banknotes and the pages of printed books.

1.33
Carlos Glidden (1834–77)

All this work was carried out at C. F. Kleinsteuber's workshop where Sholes associated with many other would-be inventors, all trying to tinker about with new and better implements. Sholes and Glidden set to work on the production of a 'Writing Machine' as a further development of their page-numbering machine. They had read about John Pratt's machine and about earlier unsuccessful attempts. The first machine they made printed only the letter 'W' but it had the basis of the right kind of mechanism.

In 1867, after many trials and errors, Sholes sent for his friends and sat down at the first machine and wrote 'C. LATHAM SHOLES SEPTEMBER 1867'. It was all in capital letters because the machine had no lower-case characters.

Sholes had learned a great deal in the building of this machine and he knew he could do better. What he did not realize was that after six years it would still be full of almost uncontrollable gremlins! 'The machine' lacked a name. They tried 'Writing Machine' and 'Printing Machine' but neither made it clear what exactly the gadget did. Sholes thought of

'Type-writer'. There was a disagreement on this point but by September 'Type-writer' was its name. Mechanical problems were immense, but the lack of money was an even larger obstacle. Manufacturers and other people liked the machine but were not willing to go any further than that, and although the inventors themselves were still very enthusiastic, they had to eat and live, and it became more and more obvious that a large amount of capital was required before they could proceed any further.

1.34
James Densmore (1820–89)

Sholes began writing letters to people offering to let them in on the ground floor in return for ready money. One of these was James Densmore of Meadville, Pennsylvania, whom Sholes had met some twenty-three years earlier and to whom he stated later he had taken an instant dislike. Densmore had been a newspaperman and a printer all his life. He was an enormous man with a huge shaggy beard who made a great deal of noise and threw his considerable weight about at every available opportunity. He was full of compulsive energy, and drove his men almost to the point of exhaustion and despair. Even when he seemed to be without funds he was apparently able to find ways of pumping money into his work. For a 25 per cent interest in the shares of the company, Densmore was to pay all the back bills amounting to around $600, and to provide all the future financing.

Densmore honoured the agreement in an impressive manner by immediately delivering $600. Poor Mr. Sholes had no idea that $600 was at that time Densmore's total funds!

People who knew him said that he always looked terrible whether he had money or not. He wore battered hats, a long shabby coat, and off-colour vests. His trousers were inches too short and he wore thick wool socks in low-cut shoes. It may be that he adopted this pose as a so-called promoter—to quote the words of Mae West: 'it is better to be looked over than overlooked'! There was no likelihood of this, however, because he seemed to frighten or upset everyone with whom he came into contact. You could almost say of him that he was looking for a fight! He was a fearless man, prepared to meet any challenge, and overcome any obstacle. There is no doubt at all that without his drive Sholes would never have been able to bring his typewriter to the production line.

He supplied tremendous optimism. He was most hopeful, even in the darkest hours when everyone else, including Sholes, was ready to give up.

Their early experimental models were given to professional writers like Roby and Weller with instructions to try and give them the toughest practical tests they could devise to show up any faults. They found plenty of these. The typebars jammed. The weight on the clock-work device that moved the carriage was too light and the string that held it was liable to break, to the jeopardy of the operator's toes. The thing stuttered and jumped. The hand-inked ribbon was a mess.

1.35 **Sholes's Experimental Keyboard of 1867; this only printed one letter**

1.36 **The Sholes, Glidden, and Soule Experimental Model of 1868 tested by Charles Weller**

1.37 Sholes's Machine Shop, Milwaukee, U.S.A. A diorama displayed in the Milwaukee Public Museum

In 1870 Sholes wrote to his friend, Weller, about his latest model:

> '. . . I think the machine is now as perfect in its mechanism as I know how to make it, or to have it made. . . . The machine is done, and I want some more worlds to conquer. Life will be most flat, stale and unprofitable without something to invent.'

but this was only Sholes's opinion. A year later he was working sixteen hours a day on a completely new model! But in all fairness, none of the workshops were suitable for the manufacture of typewriters—every part had to be made by hand which was a very laborious business. It was like trying to make watches in a blacksmith's shop. The tools were inadequate yet he carried on, first in a battered old stone building beside a canal in West Milwaukee, and then in rented premises in Chicago where fifteen typewriters, all the same design, were actually turned out. Unfortunately, the landlord wanted his money and the poor inventors were forced to retreat to Milwaukee. All the money they could acquire was spent before it was even received.

Few of the typewriters were on sale because most of those produced had either been given away for publicity purposes, or were undergoing tests and trials, or worse still, were held by creditors. Everyone was discouraged except Densmore who demanded an improved typewriter! It was said at this time that Densmore was forced to live on a diet of raw apples and biscuits eaten secretly in hotels, although this might be an exaggeration. Yet not for one fraction of a second did Densmore's faith ever waver that he could make a fortune out of the typewriter. Densmore tried to interest the Western Union and demanded $50,000 for the exclusive manufacturing rights. Western Union rejected his offer. They said they liked it, but one of their employees, Thomas A. Edison, thought that he could put something together

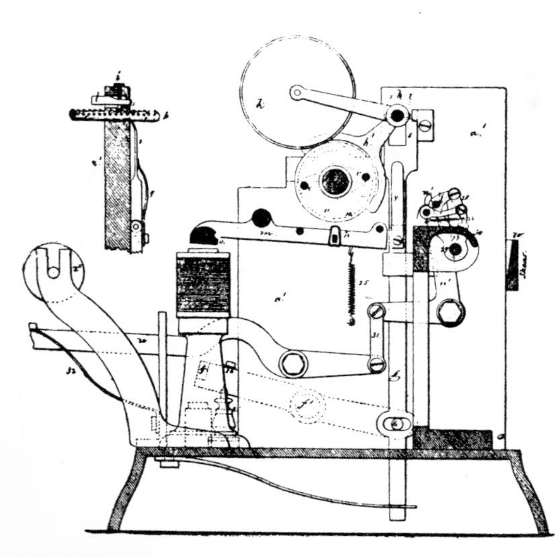

1.38 Thomas A. Edison patent drawing of 1872

1.39 The Sholes and Glidden Experimental Model of about 1873

1.40 Sholes's model of about 1873 which was originally sold to the Remington Company

for a lot less than that, and a better machine too. Western Union agreed with him.

Edison did eventually manufacture an electric printing wheel patented in 1872, but this was the forerunner of the Stock Market ticker-tape, not a typewriter.

Gradually, by either borrowing or in some way acquiring money, Densmore paid the most urgent bills. The original shareholders, seeing that they would be unlikely ever to see their money back, handed over their shares to Densmore.

Sholes still held some rights and now on the scene comes George Washington Newton Yost who had purchased some of Densmore's interest in the project.

Yost and Densmore talked Sholes into selling his rights for a flat fee. The price is not really known but was probably around $6,000 payable in instalments. It was absurdly small for all that Sholes had accomplished.

Familiar names now begin to emerge. As the story unfolds it will be seen that Densmore, Yost, and later Smith, went their own ways, eventually creating their own 'fortunes'.

Yost was one of the fastest-talking, smoothest salesmen in the world, and completely different from Densmore. He was a diplomatic person, suave, magnetic, and well-liked, and seemed far above the sordid matter of money. He had invented a reaping and sowing machine and owned a fairly large factory which he sold to raise money to go into the typewriter business.

The 1873 Sholes and Glidden typewriter which Densmore and Yost owned, bore only a faint resemblance to the first machines. It had a four-bank keyboard and the biggest difference between this machine and the modern typewriter was the under-strike idea. The printing point was invisible. Together, they tried every possible outlet. They tried to interest the Government with their invention or any department of the Government. Many were friendly and congratulated them, but no-one bought any of their machines. Densmore was infuriated!

1.41 Philo Remington (1816–89), manufacturer of the first commercial typewriter

1.42 The Sholes and Glidden machine of 1873; this was the first commercial machine and approximately 1,000 were produced. Until 1878 it only wrote capital letters. The photograph to the right shows an early user

Densmore was certain that Government offices would later teem with tens of thousands of typewriters.

Then their fortunes changed. They met Philo Remington, President of the Remington factory which was making guns, sewing-machines, and farm machinery. Remington had a magnificent factory in New York equipped with the right kind of machinery and tools that C. F. Kleinsteuber had never even dreamed of, and they were looking for something to manufacture as things were a little on the slow side. Now that the Civil War was over and most of the Indians had been shot, there was very little demand for guns!

On 1 March 1873, they signed a contract with Remington who utilized an entire wing of their plant for production of the typewriter and agreed to manufacture 1,000 machines.

Densmore and Yost were given the agency for the product and it was to bear the Remington name. Remington immediately put two of their best mechanics onto the job—William K. Jenne and Jefferson M. Clough, who together were given the difficult task of not only re-designing the machine, but also making it suitable for quantity production. In a very short time they had solved many of the problems that had haunted Sholes, Glidden, Densmore, Weller, and all the others.

Actual manufacture began in September, and the first shipments were made early in 1874. As the machine was made by sewing-machine men, it is not surprising that it looked more like a sewing-machine than a typewriter! But the first 1,000 typewriters were manufactured and they were called 'Type-writers' for practical purposes.

They were on the production line at last.

The World of 1873

Into what kind of a world was the typewriter launched? When the Remington typewriter of 1873 was introduced, only eight years had elapsed since the murder of Abraham Lincoln and the end of the Civil War in America. On the Continent the Franco-Prussian War of 1870–1

1.43 The Sholes and Glidden typewriter of 1875

had only just finished; in England, Queen Victoria had been on the throne for thirty-six years, and her Prime Minister, Gladstone was near the end of his first ministry.

Only three years earlier, in 1870 the Forster Education Act had been passed. This was a landmark in the history of education, for the government accepted for the first time the responsibility for ensuring that elementary education was available for all.

England had no serious rivals in the industrial field and had reached the peak of prosperity, and America and Germany were predominantly agricultural.

The worker's lot was harder than it is today. He started his working day earlier and finished later—and in conditions which would seem strange now. He could not switch on the electric light, for it was eight years before Edison invented the electric light bulb. He could not switch on the radio which was only to appear in rudimentary form in 1895. If he had a long way to go to work, he might well have gone by train, but certainly not by car as the first car with an internal combustion engine was not invented until 1886. Even the bicycle, as we know it today, was not to be seen on the streets and he would have been able to use nothing more than a 'bone-shaker'. When he arrived home in the evening there was no television—and not even the cinema to entertain him for cinemas did not appear until the closing years of the century. There was, of course, the piano which had been developed from the harpsichord in 1709. He could not have rung up any of his friends—the first telephone exchange in London being opened in 1879, but if he wanted to pass a piece of information quickly, he could avail himself of the telegraph service which had become a practical proposition in 1837.

The second half of the nineteenth century saw many new inventions and discoveries of the greatest importance whether in the field of medicine or in industry. Many of these were developed later in the early part of the twentieth century, sometimes under the stimulus of war demands. For instance although the first airships had flown in 1852, there was no aeroplane by the end of the century. Wright's first flight was in 1903 but the importance of the aeroplane was not fully realized until World War I. The rapid progress made in aviation owes much to the need to develop faster and more offensive planes in times of war.

The poor typewriter, being overshadowed by greater events, attracted very little attention when it was first exhibited. It was sandwiched in among many other inventions, yet it is now as indispensable as many of the important ones we have mentioned, and it has completely revolutionized the modern business world.

Like most other inventions, it was the work of many minds and many hands, and the remarkable thing is that the machines in use now are so much like those over a century ago.

The typewriter has been described as the only modern invention worth preserving. This is an extravagant claim, which few would take seriously. It was made perhaps by a busy chemist frustrated in his efforts to decipher a medical prescription, or a weary and harassed examiner marking a set of illegible scripts. However, many who sigh nostalgically for the more leisurely days of a bygone era and conjure up dreams of a pure and peaceful atmosphere, unpolluted by the jet or internal combustion engine, would be unwilling to dispense with their typewriter which has made illegibility inexcusable, if not impossible.

It is a complicated piece of machinery but it has not captured the imagination or affections of men. A man will enthuse about his powered-boat, become almost lyrical or abusive about his car, but seldom emotional about his typewriter—reserving his emotions perhaps for the typist!

There is a good reason for this—the typewriter is essentially utilitarian, associated with business rather than pleasure, so that in the office it generally attracts less attention than the typist.

Originally the word 'typewriter', although used by Sholes to describe his machine, was used to denote both the machine and—far more frequently—the 'young lady' who used it. This led inevitably to predictable Music Hall gags about men working with typewriters on their knees!

The typewriter remains a prosaic instrument, whose virtues are unsung, but whose sounds have at least been recorded. The composer Satie, whose ingenuity was more conspicuous than his genius, made use of the sound of clicking typewriter keys in his ballet 'Parade' in 1917.

The Emancipation of Women

When typewriters were first introduced the problem of finding someone to operate them had to be faced. Shorthand was in common use, but the advantages of combining this skill with typing were not immediately recognized. Few typing schools existed, which was hardly surprising in view of the limited financial prospects for typists.

'Twenty million young women rose to their feet and said "We will not be dictated to" and immediately became shorthand typists.'

—attributed to G. K. Chesterton

1.44 A Dictating Machine of 1890 with a typist at work; note storage batteries under the table

In America the Young Women's Christian Association was one of the first organizations to foresee the possibilities of a new career for women, and in 1881 they started a class of eight girls for training as typists. The idea spread rapidly, and it is estimated that five years later there were some 60,000 girls tapping on their typewriters in offices throughout the United States.

It was about this time that Rudyard Kipling, in the letters he wrote from America in 1887–9 to be published in two Indian journals, referred to the 'Typewriter Maiden' who earned her living rather than remain dependent on her parents.

The keys of the typewriter had opened up a new career for women and brought about remarkable social changes, involving a move away from the American farm towards the city centres.

Before the invention of the typewriter, career opportunities for women were few. Girls were employed in shops, factories, and domestic service, it is true, and those who enjoyed the comparatively rare privilege of a good education could contemplate a career in teaching or nursing. But whereas for men the nineteenth century could be described as an age of opportunity, it was often one of frustration and thwarted hopes for women. Parents of the middle and upper classes often frowned on the idea of their daughters earning their living and throughout the business world there was a prejudice against employing women, even in a clerical capacity. Like all prejudices, it was unreasonable, but it was nonetheless general.

As typewriters became universally acceptable more and more women entered the field of commerce—but not always without a struggle. Men who had previously used elegant copperplate in solicitors' offices and other professions and businesses, and who were the official clerks and writers, were reluctant to abandon their skill. Consequently it was not uncommon for manufacturers of typewriters to train women in the arts of typing and more or less 'sell' them to business houses with their machines, for there was no point in having a typewriter in the office if no one could use it.

Before he died in 1890, Sholes is quoted as saying 'I do feel I have done something for the women who have always had to work so hard. This will enable them more easily to earn a living'. He is also quoted as saying 'whatever I may have felt in the early days of the value of the typewriter, it is obviously a blessing to mankind, and especially to womankind. I am glad I had something to do with it. I builded wiser than I knew, and the world has the benefit of it'.

He certainly 'builded wiser' than he knew, for since those words were uttered countless women have earned their living as typists. Moreover many women who started in a relatively humble way as a copy typist, or shorthand typist, have thereby been able to progress and climb the ladder leading to higher administrative positions in many important business houses.

Today a wide variety of professions are open to women. They have become doctors, barristers, university professors, and even prime ministers.

This desirable state of affairs has been brought about by different contributory factors: the vast and accelerating growth of educational opportunity, and the employment of women to replace men in two world wars—often in responsible positions; but it seems undeniable that in the world of commerce and business, the first real opportunity for women came with the typewriter, and they have certainly made good use of it.

The Typewriter in Professional Life

Originally the typewriter seems to have found favour with professional men though solicitors are said to have been reluctant to use it, fearing that their clients would feel the lack of the personal touch. This is no longer true today, and all business and professions make use of the typewriter. However, there are occasions when the feeling still persists; letters expressing sympathy, on a bereavement, for instance, are seldom typed, lest they convey an impression of impersonality, which is often associated, perhaps mistakenly, with the typewriter.

Budding authors made good use of the typewriter, hoping no doubt that the clarity of their work would create a favourable impression with prospective editors and publishers.

Samuel L. Clemens, better known as Mark Twain, was probably the first author to submit a typed manuscript to his publisher. According to his autobiography this was the manuscript of *Tom Sawyer*, published in 1876, but some historians believe that Mark Twain's memory was fallible in this instance, and that the typed manuscript was in fact his *Life on the Mississippi* published in 1883.

This close up view (*see* Figure 1.45) of the first commercial model of Christopher Latham Sholes's 'typewriting machine' shows how E. Remington & Sons of Ilion, New York, developed the invention. This is the famous 'Model 1' of 1873, a so-called blind-writer since the line of typing could not be seen unless the carriage was lifted from the front. Typebars were arranged in a circle under the platen and the carriage was returned by means of a foot pedal attached to the sewing-machine base which was used on the Model 1 typewriters. The spacebar was made of wood (note chipped out portion in centre of bar). Keys also were made of wood. Common wood screws were used to fasten the heavy cast iron typewriter frame to the wooden top of the sewing-machine table. It sold for $125.

Figure 1.47 shows Tolstoy dictating—one of his novels(?)—to his daughter on the typewriter.

1.45 Mark Twain's Typewriter (Sholes's first commercial model of 1873)

BJUYT KIOP M LKJHGFDSA:QWERTYUIOP:_-98VB643234 RT
HA

HARTFORD, DEC. 9,

DEAR BROTHER:

I AM TRYING T TO GET THE HANG OF THIS NEW F
FANGLED WRITING MACHINE, BUT AM NOT MAKING
A SHINING SUCCESS OF IT. HOWEVER THIS IS THE
FIRST ATTEMPT I EVER HAVE MADE, & YET I PER-
CEIVETHAT I SHALL SOON & EASILY ACQUIRE A FINE
FACILITY IN ITS USE. I SAW THE THING IN BOS-
TON THE OTHER DAY & WAS GREATLY TAKEN WITH
IT. SUSIE HAS STRUCK THE KEYS ONCE OR TWICE,
& NO DOUBT HAS PRINTED SOME LETTERS WHICH DO
NOT BELONG WHERE SHE PUT THEM.
THE HAVING BEEN A COMPOSITOR IS LIKELY TO BE
A GREAT HELP TO ME, SINCE O NE CHIEFLY NEEDS
SWIFTNESS IN BANGING THE KEYS. THE MACHINE COSTS
125 DOLLARS. THE MACHINE HAS SEVERAL VIRTUES
I BELIEVE IT WILL PRINT FASTER THAN I CAN WRITE.
ONE MAY LEAN BACK IN HIS CHAIR & WORK IT. IT
PILES AN AWFUL STACK OF WORDS ON ONE PAGE.
IT DONT MUSS THINGS OR SCATTER INK BLOTS AROUND.
OF COURSE IT SAVES PAPER.

SUSIE IS GONE,

NOW, & I FANCY I SHALL MAKE BETTER PROGRESS.
WORKING THIS TYPE-WRITER REMINDS ME OF OLD
ROBERT BUCHANAN, WHO, YOU REMEMBER, USED TO
SET UP ARTICLES AT THE CASE WITHOUT PREVIOUS-
LY PUTTING THEM IN THE FORM OF MANUSCRIPT. I
WAS LOST IN ADMIRATION OF SUCH MARVELOUS
INTELLECTUAL CAPACITY.

LOVE TO MOLLIE.
YOUR BROTHER,
SAM.

1.46 Mark Twain's first typewritten letter dated 1874

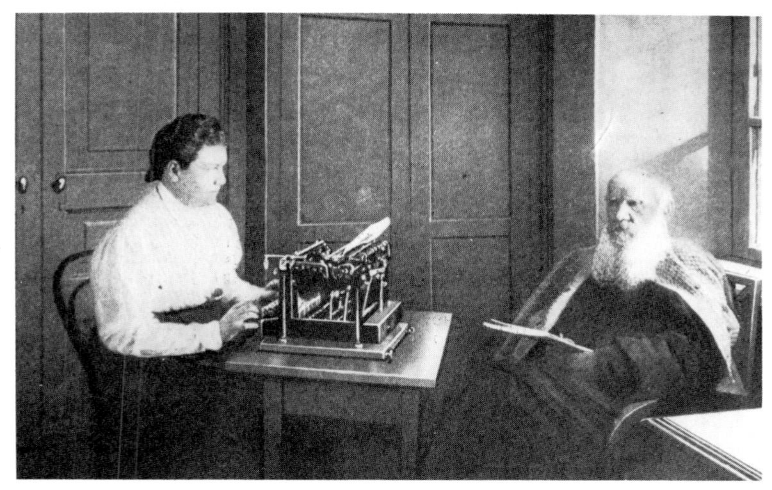

1.47 Count Tolstoy said to be dictating his novel *War and Peace* **to his daughter**

As far as is known the first part played in fiction by a typewriter is recorded in Conan Doyle's *A Case of Identity*, published in 1892. In this story Sherlock Holmes was able to solve the mystery by correctly identifying the impostor's typewriter!

The Typewriter During War

During the two World Wars it is hardly surprising that the typewriter industry took a heavy knock.

In World War I in 1914–18, typewriter production ground slowly almost to a halt. Typewriters had often begun their life in strange surroundings—Remington for instance, had been manufacturing arms, and Kleyer, the precursor of Adler, sewing-machines. When war came, it was inevitable that many of the production lines of typewriters should be converted to producing the essential munitions of war—whether guns, small arms, fuses or other precision instruments. As America entered the war at a comparatively late stage, their typewriter production suffered less than in European countries, and they were able to continue their exports for some considerable time.

The American Corona—a very light three-bank portable—was extensively used by the the British Forces during the war. Thousands of these machines saw service in the trenches, and it is also known that they were used for reporting in patrol aircraft. Some very important documents, including the surrender papers for German South West Africa, were typed and duplicated on a Corona typewriter.

In 1914, Russia had been well stocked with American and German typewriters such as the Continental, Adler, Mercedes, and Stoewer. When hostilities broke out the demand increased, but supplies became more and more difficult. Large numbers of the American 'Victor' typewriters were imported by way of Sweden, but when the Finnish railways became overcrowded, supplies of these, too, dried up.

After the war, prices for typewriters rocketed, as they had become almost unobtainable, and retailers were able to sell anything they could acquire. A similar state of affairs existed during and after World War II.

Shortly after America entered World War II, all their typewriter manufacturing facilities were converted to war, with one exception. The exception was the Woodstock Typewriter Company, which was authorized to produce up to 18,000 typewriters a year.

1.48 A Corona Typewriter being used in the field during World War I

1.49 The Typewriter used during World War I by the Canadian Light Artillery

The American Government decided to buy in typewriters from users, through manufacturers and retailers and the public was asked to release any spare typewriters. They certainly had to search high and low, for the American Forces required about 650,000—mostly for their Army—and of course other government departments needed some as well. The date when the War Production Board froze all sales of new and used typewriters was 5 March 1942. This was more than a year before the British Government clamped down on typewriter sales.

The main typewriter used by the British Army was the Oliver, and at the beginning of World War II, large numbers of these machines were used.

On 7 May 1943 the British Government issued an order controlling the supply of typewriters so that the heavy demands of the fighting services and government factories could be met. Together these absorbed nearly 80 per cent of the total output. No retailers or even private individuals were allowed to buy or sell typewriters without a licence unless the machines were second-hand portables weighing less than 22 lb.

In Germany many typewriter factories were seriously damaged in air raids, with consequent disruption not only of the production of typewriters, but also of vital munitions. The Adler factory in Germany, for instance, was manufacturing amongst other items such as precision parts for tanks, part of the rear portion of torpedos, while IBM were similarly employed in contributing to the manufacture of torpedos in the United States. American and British typewriter factories fortunately remained intact, and typewriter losses were sustained only in individual offices and industrial premises which had been bombed, or else in a few special cases, such as the unfortunate incident when, a few days after the landing in Normandy, a ship loaded with typewriters was mined and sunk. Twenty thousand Underwood and Royal machines were lost.

CHAPTER 2

The Typewriter Today

THE STRUGGLE FOR THE CORRECT KEYBOARD

All typewriters produced today, with certain well-established exceptions such as the IBM '72/82 Golf Ball', are front-strike machines with segment and visible writing, produced either through a fabric or carbon ribbon, with a shift key to move into the capital position. This applies to electric, standard, and portable models.

When the early producers of typewriters first directed their thoughts to a keyboard, they were obsessed with the arrangement of the piano keyboard. They just could not see how anything better could be conceived. It was generally believed that if a keyboard of eight letters (like a scale on a piano) could be extended to twenty-six letters, it would have been a perfectly simple arrangement and that, had the 'piano' typewriter been put into production, a vast number of people could easily be trained very quickly to use a machine without any difficulty. This was probably very sound thinking.

People, after all, had been playing pianos for 200 years and remember, the basic principles of the piano have changed very little, and the keyboard remains the same today. It is universally understood in any country throughout the world. Given these simple facts, perhaps it is understandable that those who were striving to make a Writing Machine could not see beyond this musical instrument and its general layout. The Francis typewriter was a good example of this.

Early machines showed a vast variety of keyboard arrangements. Some were circular, others had three to eight or ten rows of keys; and some had no shift keys whilst others had one or two. Originally the tendency was to arrange all letters, both uppercase (capitals) and lowercase (small letters) in alphabetical order for easy reference. It was assumed that if people knew their alphabet, and most people did, it would be easy to locate the letter required. It was also assumed that where numbers were required, these would be in numerical order as, indeed, they still are, but Sholes who in 1873 produced the first production typewriter, soon ran into difficulties. He found that the 'ABC' arrangement caused his up-strike machine to jam when any speed was reached and, realizing the insurmountable technical problems arising from this, which had exhausted both his skill and patience, he cast around for other means of resolving his dilemma. He sought the advice of his brother-in-law who was a schoolmaster and mathematician, and asked him to re-arrange the keyboard so that, on most occasions, the bars would come up from opposite directions and would not clash together and jam the machine.

After many calculations and experiments, Sholes established the existing keyboard on which the first six letters are Q W E R T Y, and departed from all previous alphabetical arrangements. He then proceeded to sell this 'QWERTY' arrangement of the keyboard. It was probably one of the biggest confidence tricks of all time—namely the idea that this arrangement of the keyboard was scientific and added speed and efficiency. This, of course, was true of his particular machine, but the idea that the so-called 'scientific arrangement' of the keys was designed to give the minimum movement of the hands was, in fact, completely false! To write almost any word in the English language, a maximum distance has to be covered by the fingers.

It has been claimed that any haphazard arrangement of letters would be mathematically better than the existing one. However, Sholes's 'Remington QWERTY' keyboard became well established and nobody seemed to question the accuracy of Sholes's statement. All other manufacturers adopted the arrangement with only slight variations. Those who had double shift three-bank keyboards, arranged their letters in this manner, and those who had gigantic double keyboards did the same. Those who failed to do so, disappeared without any trace. Hammond, for instance, tried to challenge this arrangement but had to withdraw in disorder and convert to 'QWERTY'.

But whether manufacturers should think in terms of three-bank, four-bank or double keyboards, was decided in a curious and dramatic way.

A certain Mrs. L. B. Longley, the owner of a Shorthand and Typing Institute, had the audacity to state that all typists should use all the fingers of both hands. The *Cosmopolitan Shorthander* in 1877 condemned Mrs. Longley and stated that unless the third finger of the hand had been previously trained to touch the keys of a piano, it was not worth while attempting to use this finger in operating the typewriter. It went on to say that the best operators all used only the first two fingers of each hand and doubted whether a higher speed could be obtained by the use of three. Such an important publication could have put Mrs. Longley in her place had it not been for Mr. Frank E. McGurrin of Salt Lake City who, quite accidentally, rescued Mrs. Longley and established the four-bank keyboard once and for all.

He was the stenographer for the Federal Court in Salt Lake City and a first class typist. He used all ten fingers on the Remington Model '1' and he could really make that typewriter move but—and this was most important—he had memorized the keyboard! He could type blindfolded if necessary. He was proud of his achievements, which others regarded not merely as exaggerated, but impossible!

Mr. McGurrin gave demonstrations of fantastic typing ability to gasping audiences all over the West. He was a mild character until the speed of typing was mentioned when he always proclaimed in no uncertain manner that he was the fastest typist in the world! Furthermore, he was prepared to challenge anyone who had the audacity to suggest that he was not. He was ready to take on anyone at any time, in any place, and he was ready to bet a substantial sum of money he would win.

Now there was a certain Mr. Louis Taub who thought poorly of ten-finger typists. He believed four fingers were enough. Moreover, he thought it was ridiculous to type on only four rows of keys and was a firm believer in the 'Caligraph' and the double keyboard arrangement. He was *certain* that *he* was the fastest typist in the world! Mr. McGurrin was furious and flung down the gauntlet! Mr. Taub accepted the challenge. They travelled to Cincinatti for the 'duel'. The world press was attracted. It was 'News'! It was something like an international chess tournament, and took place in two parts—forty-five minutes of copying from an unfamiliar script, and forty-five minutes dictation. The one with the largest combined total number of words would win and would take $500. Never before had a duel been fought with typewriters! The event stirred up a great deal of interest—far more than the contestants had ever expected, and poor Mrs. Longley was in a dreadful state surrounded by enormous publicity.

Suddenly, to their horror, it dawned upon both the Remington Company and the Caligraph Company officials, torn between pride and despair, that whoever won was likely to put the other out of business!

Mr. McGurrin won easily, as he had predicted. He won both events and therefore the aggregate. The duel was widely reported all over the world. Mr. McGurrin had actually moved faster when he was working from copy than when he had taken dictation. His eyes were glued to the material, never losing his place, and never looking at his hands, whilst poor Mr. Taub could only take in an eyeful at a time. He fell further and further behind, wagging his head from side to side like a spectator at a tennis match.

This most unusual duel was curiously decisive. It was immediately clear to everyone, and particularly to Mr. Taub, that a good four-finger man did not stand a chance against a good ten-finger man. The keyboard had to be memorized and thus, inevitably, the double keyboard was doomed, because it was much too large to negotiate by touch and memory alone.

The Remington Company and all exponents of four-bank machines were delighted, but the poor Caligraph Company and the other double keyboard people had to admit they were wrong and if they wanted to survive at all, they either had to produce a four-bank machine with a shift key, get out of the business as fast as possible or start making something else. Furthermore, there were few typists prepared to say they were faster than Mr. McGurrin—at least not anywhere he was likely to hear or read about it!

In 1905 a large international meeting was called to establish a standard keyboard once and for all. At that time various keyboards—certainly more efficient than the one devised by Sholes and used today—were put forward as alternatives. The battle raged backwards and forwards. Nobody could agree on what a new keyboard should be, but the biggest opposition came from *teachers of typing* as it still does today. They wanted things to remain as they were, and they are still most reluctant to change their methods and learn all over again.

All present keyboards are, therefore, based on the 'QWERTY' layout.

At the end of the first 100 years of the history of the typewriter, we are once again assailed by those who wish to change—talk, talk, talk—letters to manufacturers, but nothing achieved.

It has been pointed out that many manual and electric machines in use could operate much faster but for the limitations imposed by the inefficient arrangement of the keyboard as we now have it. Manufacturers are prepared to produce different keyboards at no extra cost, as it is only a question of which letter goes on which typebar.

In this context, it is interesting to note that in most European languages, the letter 'e' is used about 200 times as often as the letter 'z'. Some letters are rarely used at all. Many studies of the frequency of letters in the English language have been made and the general opinion is as follows:

E T A O N R I S H D L F C
M U G Y P W B V K X J Q Z

A E I O U, the vowels, make up 39 per cent of all letters used and the five most common consonants are:

H N R S T

In the English language, the most common two-letter words are:

OF	TO	IN	IS	IT
BE	AS	AT	SO	WE
HE	BY	OR	ON	DO
IF	ME	MY	UP	AN

The most common three-letter words are:

THE	AND	FOR	ARE	BUT
NOT	YOU	ALL	ANY	CAN
HAD	HER	WAS	ONE	OUR
OUT	DAY	GET	HAS	HIM

The most common four-letter words are:

THAT	WITH	HAVE	THIS	WILL
YOUR	FROM	THEY	KNOW	WANT
BEEN	GOOD	MUCH	SOME	TIME
VERY	WHEN	COME	HERE	JUST

The most common reversed letters are:

ER and RE
ES and SE
AN and NA
TI and IT
ON and NO

The most common double letters are:

LL	EE	SS	OO
TT	FF	RR	NN
PP	CC		

Any typewriter factory could, without any delay or interruption of production lines, start producing a typewriter based on the above frequency pattern, or any other system.

In fact Dr. August Dvorak, who was Professor of Education and Director of Research at the University of Washington in Seattle, devised in 1932, a simplified keyboard, which he claimed would accelerate the speed of typing by about 35 per cent. His keyboard arrangement known as 'DSK' is shown in Figure 2.1 together with the usual 'QWERTY' arrangement (Figure 2.2).

But the simple fact remains that no one buys, or wants these simplified keyboards in spite of their obvious advantages. Probably in the year 2005 there will be another conference and the chances are that those who train and teach will be opposed to any change in the keyboard for, as Dr. Dvorak said, proposing a new typewriter keyboard was like proposing to 'Reverse the Ten Commandments and the Golden Rule, discard every moral principle, and ridicule motherhood'!

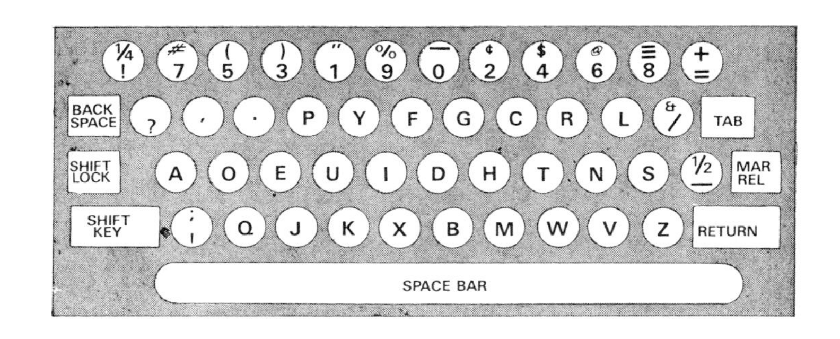

2.1 Dr. August Dvorak's simplified keyboard

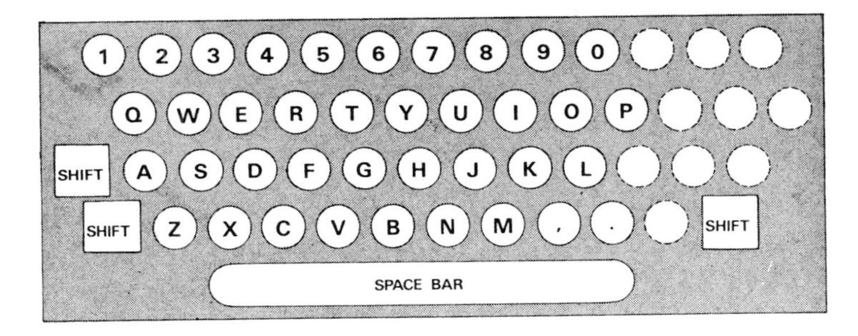

2.2 The 'QWERTY' keyboard

KEYBOARDS FOR ALL THE WORLD

In the following pages are illustrated firstly, the early keyboards and variations on them (2.3–2.9); then the generally accepted keyboard layouts for individual countries (2.10–2.286), followed by special keyboards and symbols for special purposes (2.287–2.302). The new proposed international keyboard and the variations or type combinations which are available for various professions and industries as additions or substitutions are included.

Some older machines and some lower priced portable typewriters have forty-two keytops typing eighty-four characters by double shift. Other machines have forty-six keytops writing ninety-two characters. This, of course, leads to variations in keyboards but those shown are, generally speaking, standard keyboards accepted in various countries. Certain languages such as Russian and Greek, have two keyboards, and with Arabic and Hebrew machines, the carriage moves in the opposite direction.

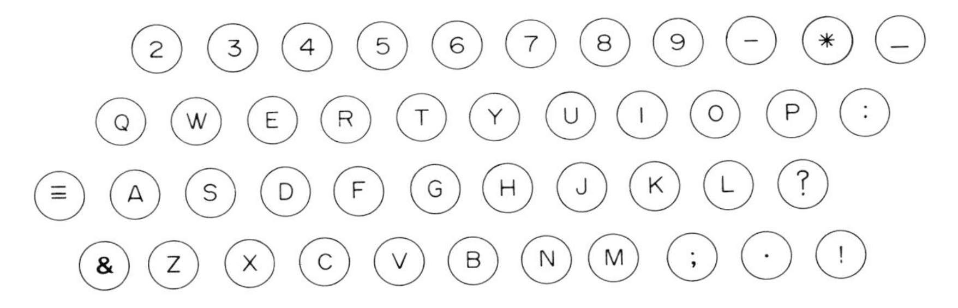

2.3 The original Sholes keyboard, 1873

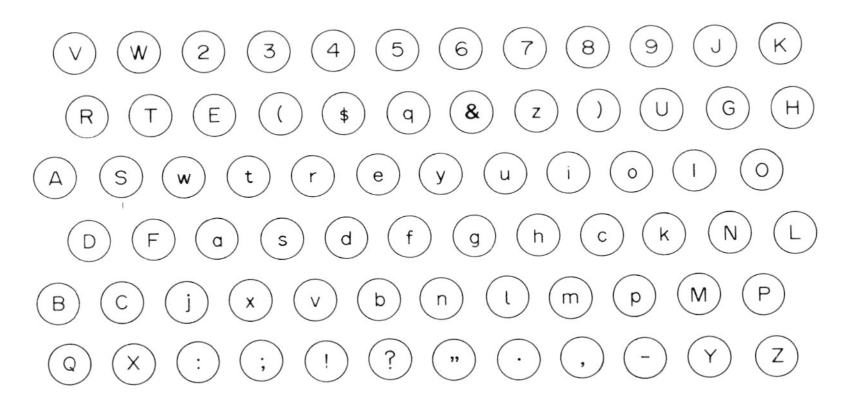

2.4 Caligraph keyboard

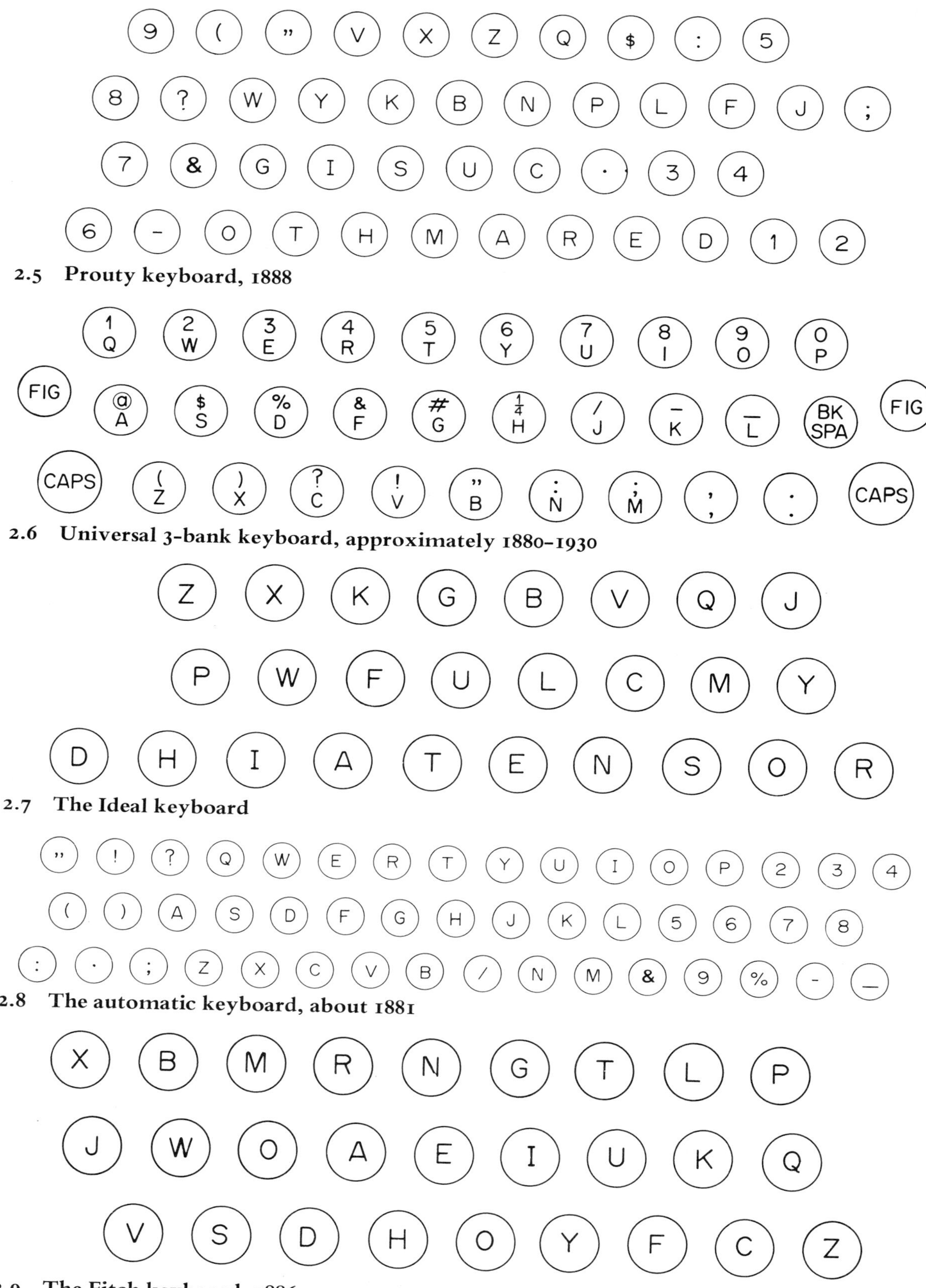

2.5 Prouty keyboard, 1888

2.6 Universal 3-bank keyboard, approximately 1880–1930

2.7 The Ideal keyboard

2.8 The automatic keyboard, about 1881

2.9 The Fitch keyboard, 1886

2.10–2.286 Complete range of modern keyboards (by courtesy of Olympia International)

See page 43

2.10 Afrikaans

2.11 Afrikaans

2.12 Afrikaans

2.13 Afrikaans 2

2.14 American

2.15 American

2.16 American

2.17 American 9

2.18 American 6

2.23 American 8

2.19 American 6

2.24 American ASA

2.20 American 7

2.25 American 17

2.21 American 7

2.26 American 17

2.22 American 8

2.27 American 17

2.28 American 20

2.33 American-Med/Drug

2.29 American 21

2.34 American-Med/Drug

2.30 American-Chemistry

2.35 Amharic

2.31 American-Chemistry

2.36 American Air Force

2.32 American 5 ANSI

2.37 Anglo-Iraq

2.38 Anglo-Israel

2.43 Anglo-Persian

2.39 Anglo-Israel

2.44 Anglo-Persian

2.40 Anglo-Japan

2.45 Anglo-Thailand

2.41 Anglo-Japan

2.46 Anglo-Siamese

2.42 Anglo-Japan

2.47 Anglo-Universal

2.48 Anglo-Universal

2.49 Anglo-Universal 1

2.50 Anglo-Universal 2

2.51 Anglo-Universal 2

2.52 Anglo-Universal 3

2.53 Anglo-Universal 3

2.54 Anglo-Universal 4

2.55 Anglo-Universal 4

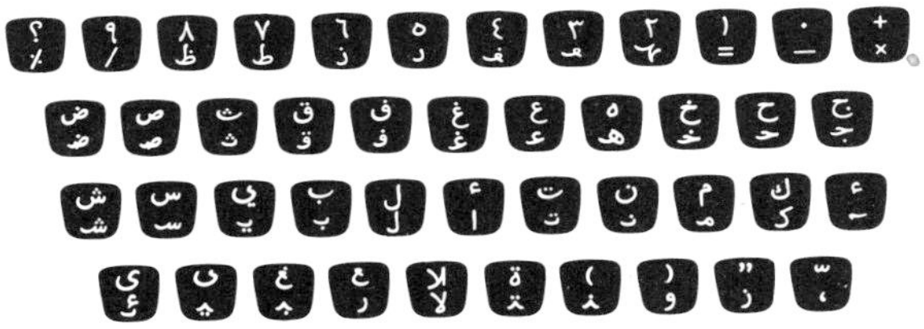

2.56 Arabic

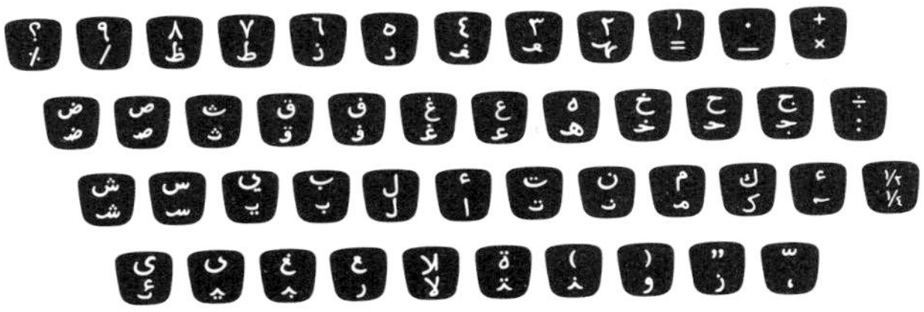

2.57 Arabic

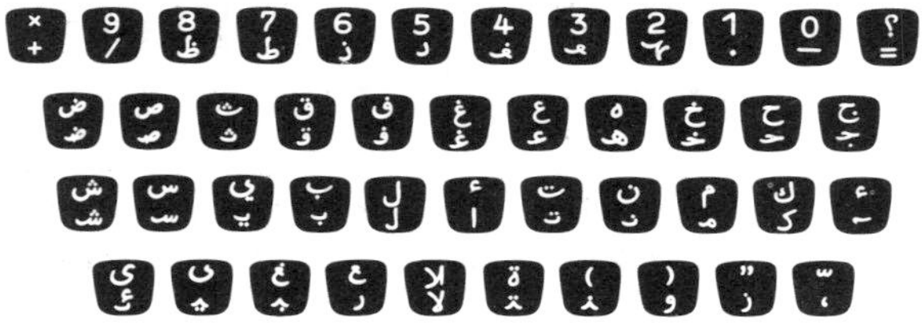

2.58 Arabic 3

2.63 Belgian

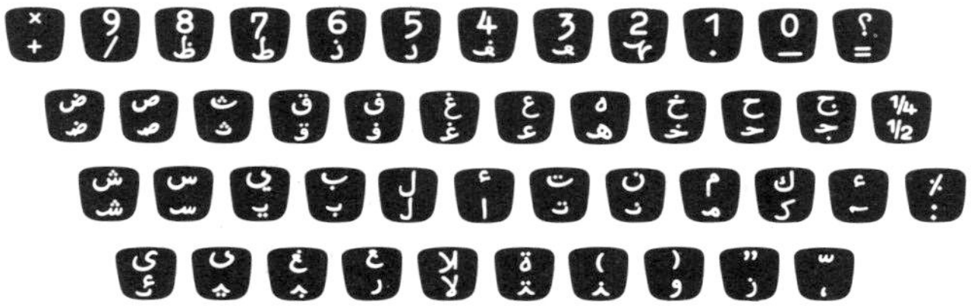

2.59 Arabic for Morocco

2.64 Bolivia 5

2.60 Armenian 2 RO 340

2.65 Bolivia 5

2.61 Armenian 2 Ro 340

2.66 Brazilian

2.62 Belgian

2.67 Brazilian 3

1 2 3 4 5 6 7 8 9 Ç $ –
Q W E R T Y U I O P `
A S D F G H J K L Ç ^
Z X C V B N M , . a

2.68 **Brazilian 4**

1 2 3 4 5 6 7 8 9 0 I V
Ф У Е И Ш Щ К С Д З Ц
Ь Я А О Ж Г Т Н В М Ч
Ю Й Ъ Ы Э Х П Р Л Б

2.69 **Bulgarian**

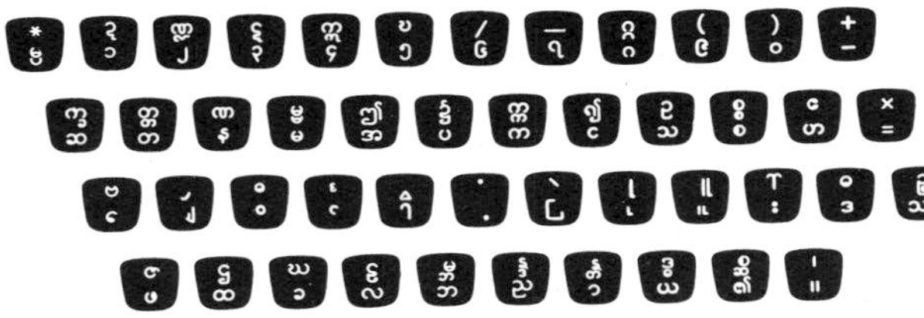

/ № 1 2 4 5 6 7 8 9 I V
Ф У Е И Ш Щ К С Д З Ц
Ь Я А О Ж Г Т Н В М Ч
Ю Й Ъ Ы Э Х П Р Л Б

2.70 **Bulgarian 2**

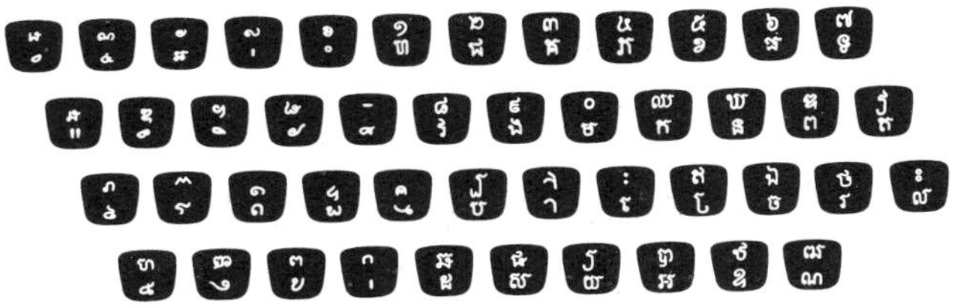

/ № 1 2 4 5 6 7 8 9 I V
Ф У Е И Ш Щ К С Д З Ц ;
Ь Я А О Ж Г Т Н В М Ч)
Ю Й Ъ Ы Э Х П Р Л Б

2.71 **Bulgarian 2**

2.72 **Burmese**

2.73 **Burmese**

2.74 **Burmese 2**

2.75 **Cambodia 2**

1 2 3 4 5 6 7 8 9 0 – =
Q W E R T Y U I O P ½ th
A S D F G H J K L ; ¢ ¾
Z X C V B N M , . /

2.76 **Canada-English**

1 2 3 4 5 6 7 8 9 0 – !
Q W E R T Y U I O P ½
A S D F G H J K L ; ^
Z X C V B N M , . ç

2.77 **Canada-French**

2.78 **Canada-French**

2.79 **Canada-French**

2.80 **Chile 6**

2.81 **Chile 6**

2.82 **Czechoslovakian Norm.**

2.83 **Czechoslovakian Norm.**

2.84 **Danish**

2.85 **Danish**

2.86 **Danish 2**

2.87 **Danish 5**

2.88 **Danish 5**

2.93 **Dutch 4**

2.89 **Dominican 2**

2.94 **Dutch 5**

2.90 **Dominican 2**

2.95 **Dutch Antilles**

2.91 **Dutch 1**

2.96 **Dutch Antilles**

2.92 **Dutch 3**

2.97 **Egyptian (French for Egypt)**

2.98 Egyptian (French for Egypt)

2.103 English 7

2.99 English for Egypt

2.104 English 7

2.100 English 3

2.105 English for Malta

2.101 English 4

2.106 English for Malta

2.102 English 7

2.107 Finnish-Standard

2.108 **Finnish-Standard**

2.113 **French 7**

2.109 **Finnish-Standard**

2.114 **French 8**

2.110 **French 2**

2.115 **French 9**

2.111 **French 2**

2.116 **French 9**

2.112 **French 3**

2.117 **French 4-ASA**

2.118 **French 5-ASA**

2.123 **German-French**

2.119 **French for Equatorial Africa**

2.124 **German-French**

2.120 **German-DIN**

2.125 **German-French 2**

2.121 **German-DIN**

2.126 **German-American**

2.122 **German**

2.127 **German-DIN**

2.128 Greek

2.133 Greek-Latin 5

2.129 Greek

2.134 Greek-Latin 6

2.130 Greek-French 2

2.135 Guatemala 3

2.131 Greek-French 3

2.136 Gujrati

2.132 Greek-French 4

2.137 Gujrati 2

2.138 **Hebrew 5**

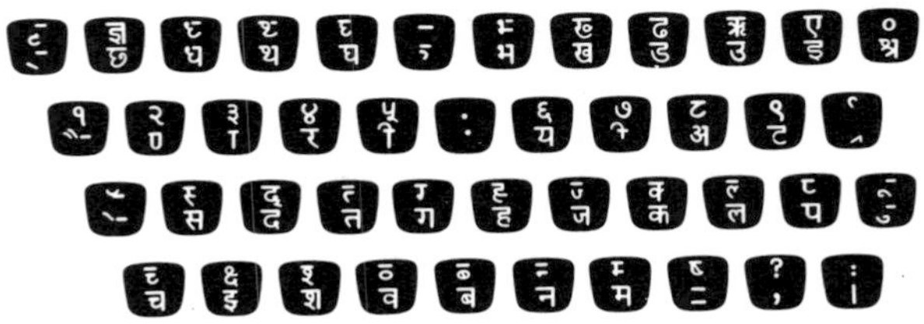

2.143 **Hindi for UP**

2.139 **Hebrew 5**

2.144 **Hindi for UP**

2.140 **Hindi 2**

2.145 **Honduras**

2.141 **Hindi FL**

2.146 **Honduras**

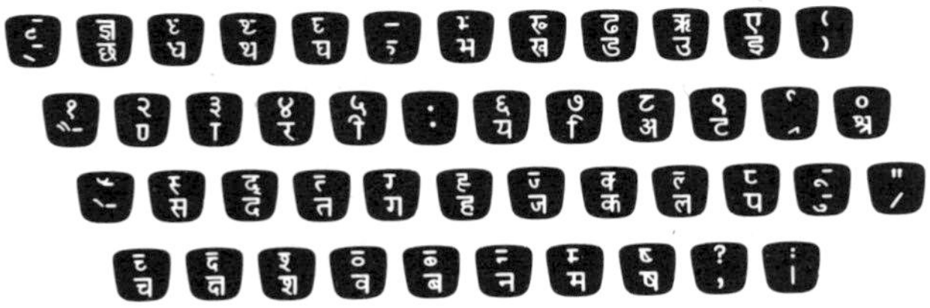

2.142 **Hindi for Bihar**

2.147 **Hong Kong Government**

2.148 **Hungarian**

2.149 **Hungarian**

2.150 **Hungarian 2**

2.151 **Iceland 3**

2.152 **Iceland 3**

2.153 **Indonesian**

2.154 **Indonesian**

2.155 **Italian**

2.156 **Italian 5**

2.157 **Italian 2**

2.158 Italian 3

2.163 Katakana-Latin 4

2.159 Japan/Farrington

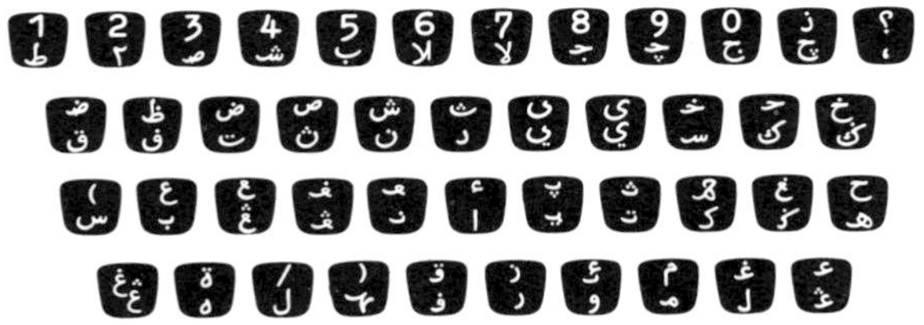

2.164 Katakana-Latin 4

2.160 Java

2.165 Korea

2.161 Katakana Japan

2.166 Korea 2

2.162 Katakana Japan 3

2.167 Laos

2.168 Lettish 3

2.169 Library

2.170 Lithuania 5

2.171 Luxembourg

2.172 Luxembourg

2.173 Macedonian-Cyrillic

2.174 Macedonian-Cyrillic

2.175 Malta

2.176 Misra-Hindi

2.177 Mozambique-Angola 2

2.178 Mozambique-Angola 2

2.183 Nicaragua

2.179 Mozambique-Angola 2

2.184 Norwegian Standard

2.180 New Zealand

2.185 Norwegian Standard

2.181 New Zealand

2.186 Norwegian 2

2.182 Nicaragua

2.187 Norwegian

2.188 **Panama 3**

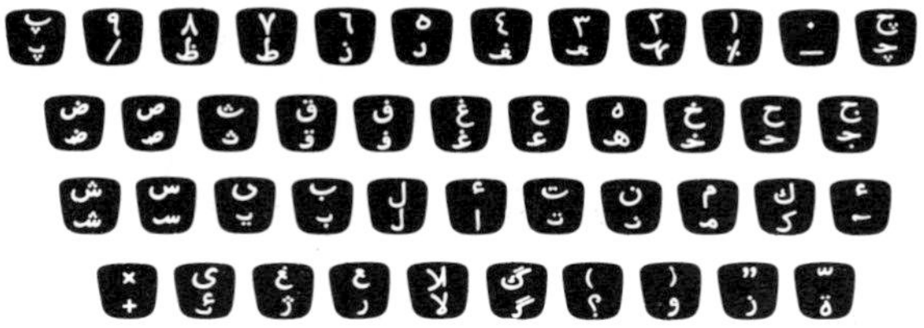

2.193 **Persian**

2.189 **Panama 3**

2.194 **Persian 2**

2.190 **Paraguay 2**

2.195 **Philippines 2**

2.191 **Paraguay 2**

2.196 **Philippines 2**

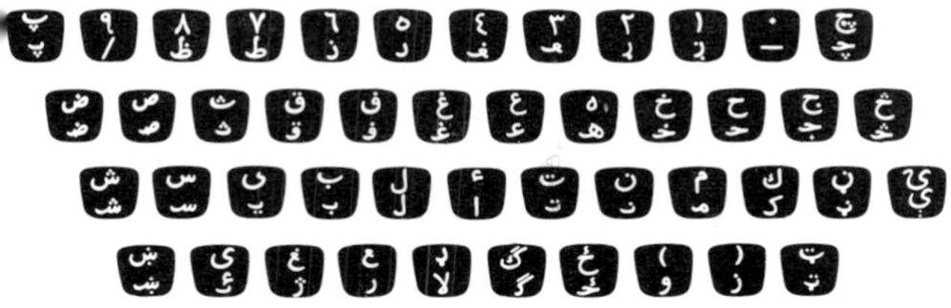

2.192 **Pashtu**

2.197 **Philippines 2**

2.198 **Polish Standard**

2.199 **Polish Standard**

2.200 **Portuguese 2**

2.201 **Portuguese 3**

2.202 **Portuguese 3**

2.203 **Portuguese 4**

2.204 **Portuguese 4**

2.205 **Portuguese 5**

2.206 **Portuguese 5**

2.207 **Roumanian**

2.208 Roumanian

2.213 Russian Standard

2.209 Roumanian 2

2.214 Russian Standard 2

2.210 Russian

2.215 Russian-Ukraine

2.211 Russian

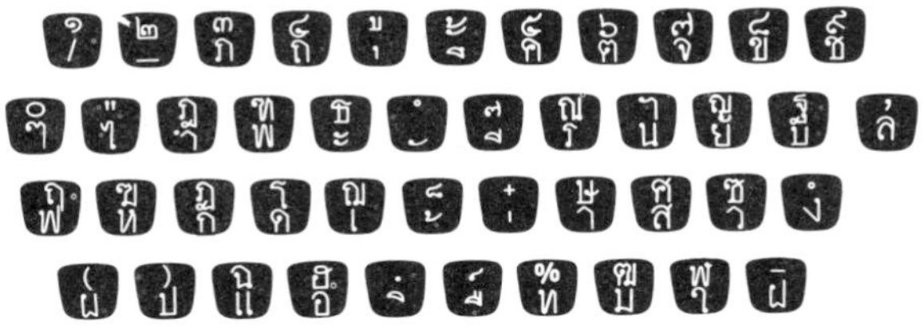

2.216 Siamese Standard

2.212 Russian Standard

2.217 Siamese Standard

2.218 **Siamese Standard 2**

2.219 **Siamese Standard 2**

2.220 **Siamese Standard 3**

2.221 **Siamese Standard 3**

2.222 **Siamese Standard 4**

2.223 **Siamese Standard 4**

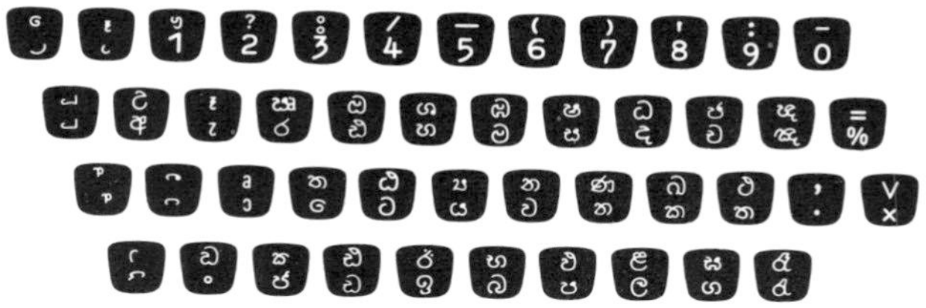

2.224 **Singhalese**

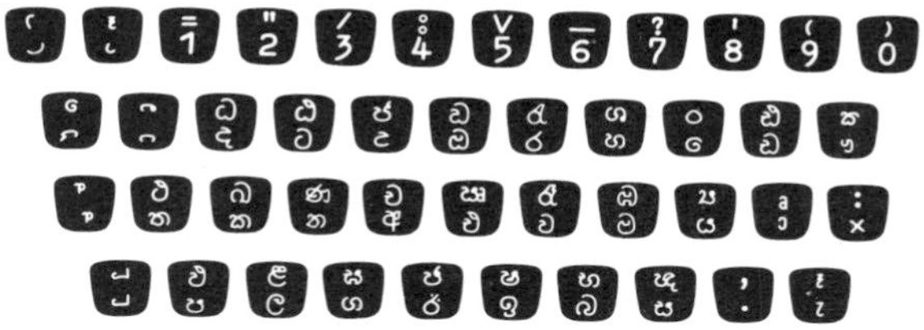

2.225 **Singhalese 7**

2.226 **Slovakian**

2.227 **Spanish**

2.228 Spanish 2

2.233 Spanish 3

2.229 Spanish 2

2.234 Spanish 6

2.230 Spanish 3

2.235 Spanish 6

2.231 Spanish 3

2.236 Spanish 5

2.232 Spanish 4

2.237 Spanish 6

2.238 Spanish 7

2.239 Spanish-Hispano

2.240 Spanish-Hispano

2.241 Sudan 2

2.242 Sudan 2

2.243 Suriname

2.244 Suriname

2.245 Swedish 2

2.246 Swedish 3

2.247 Swedish 3

2.248 **Swedish-Pharmacy**

2.249 **Swiss**

2.250 **Swiss 2**

2.251 **Swiss 2**

2.252 **Swiss 3**

2.253 **Swiss 4**

2.254 **Swiss 5**

2.255 **Swiss 6**

2.256 **Tamil**

2.257 **Tamil 2**

2.258 Telegraphy 1

2.263 Turkish 4

2.259 Telegraphy 2

2.264 Turkish 4

2.260 Telegraphy 3

2.265 Ukraine

2.261 Telegraphy 4

2.266 Ukraine

2.262 Telegraphy 4

2.267 Universal

2.268 Universal

2.273 Universal 6

2.269 Universal 2

2.274 Universal 7

2.270 Universal 4

2.275 Universal 8

2.271 Universal 5

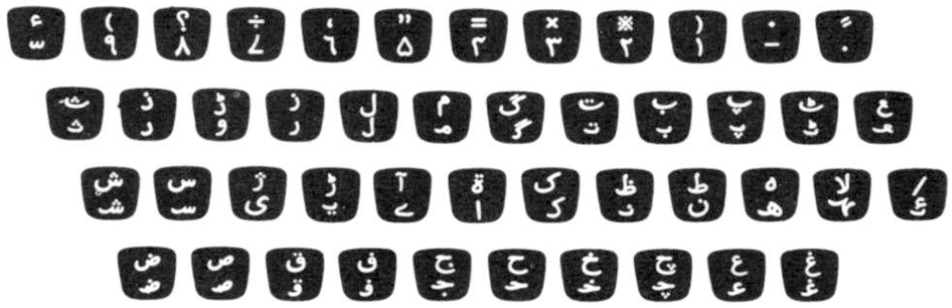

2.276 Urdu 2

2.272 Universal 5

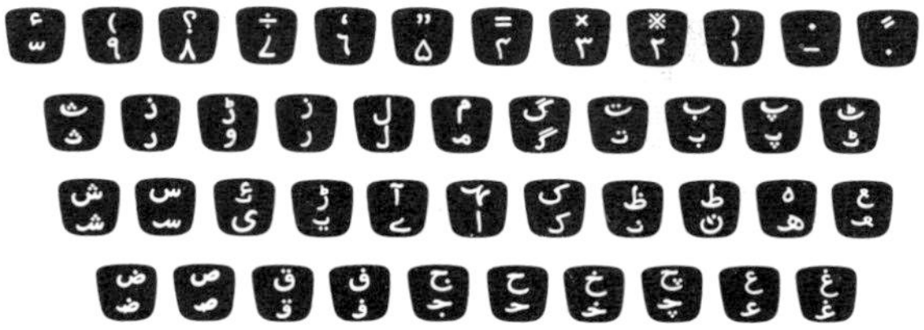

2.277 Urdu 5

2.278 Venezuela 5

2.279 Venezuela 6

2.280 Vietnam 3

2.281 Vietnam 3

2.282 Yugoslavia

2.283 Yugoslavia 2

2.284 Yugoslavia 2

2.285 Yugoslavia-Cyrillic

2.286 Yugoslavia-Cyrillic

Note: In the special symbols the numerals below the keys are the Olympia International reference.

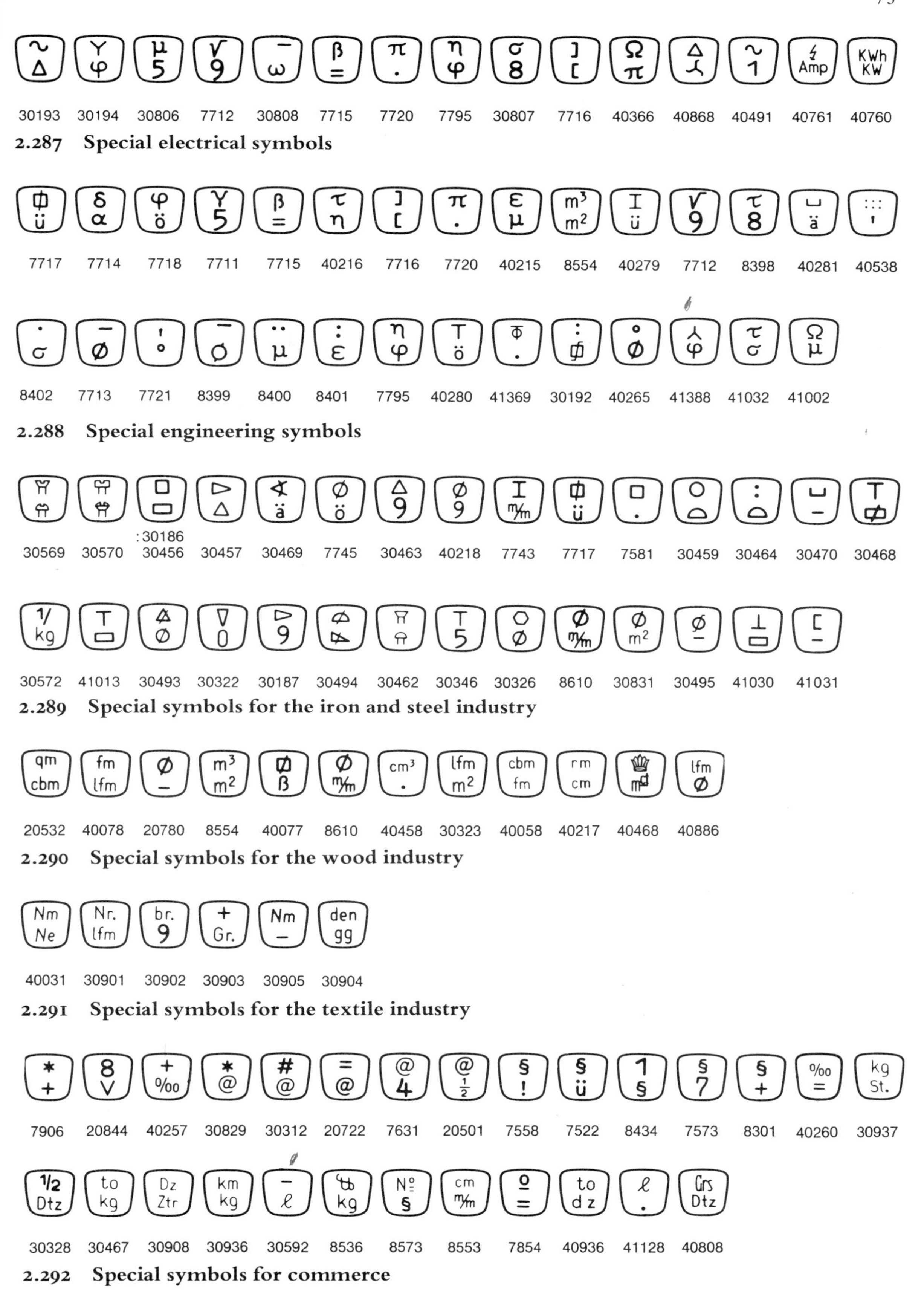

2.287 Special electrical symbols

2.288 Special engineering symbols

2.289 Special symbols for the iron and steel industry

2.290 Special symbols for the wood industry

2.291 Special symbols for the textile industry

2.292 Special symbols for commerce

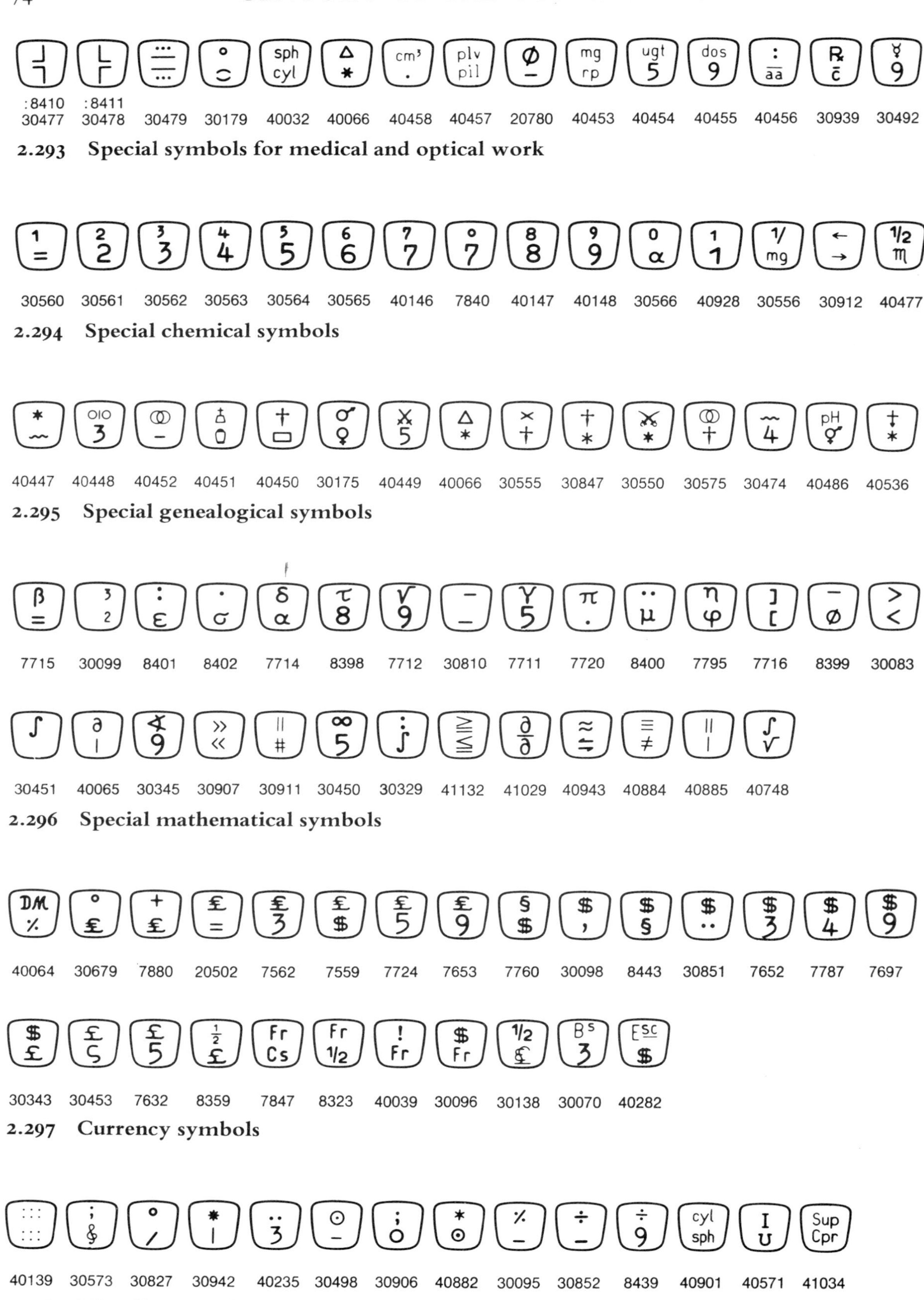

2.293 Special symbols for medical and optical work

2.294 Special chemical symbols

2.295 Special genealogical symbols

2.296 Special mathematical symbols

2.297 Currency symbols

2.298 Miscellaneous signs

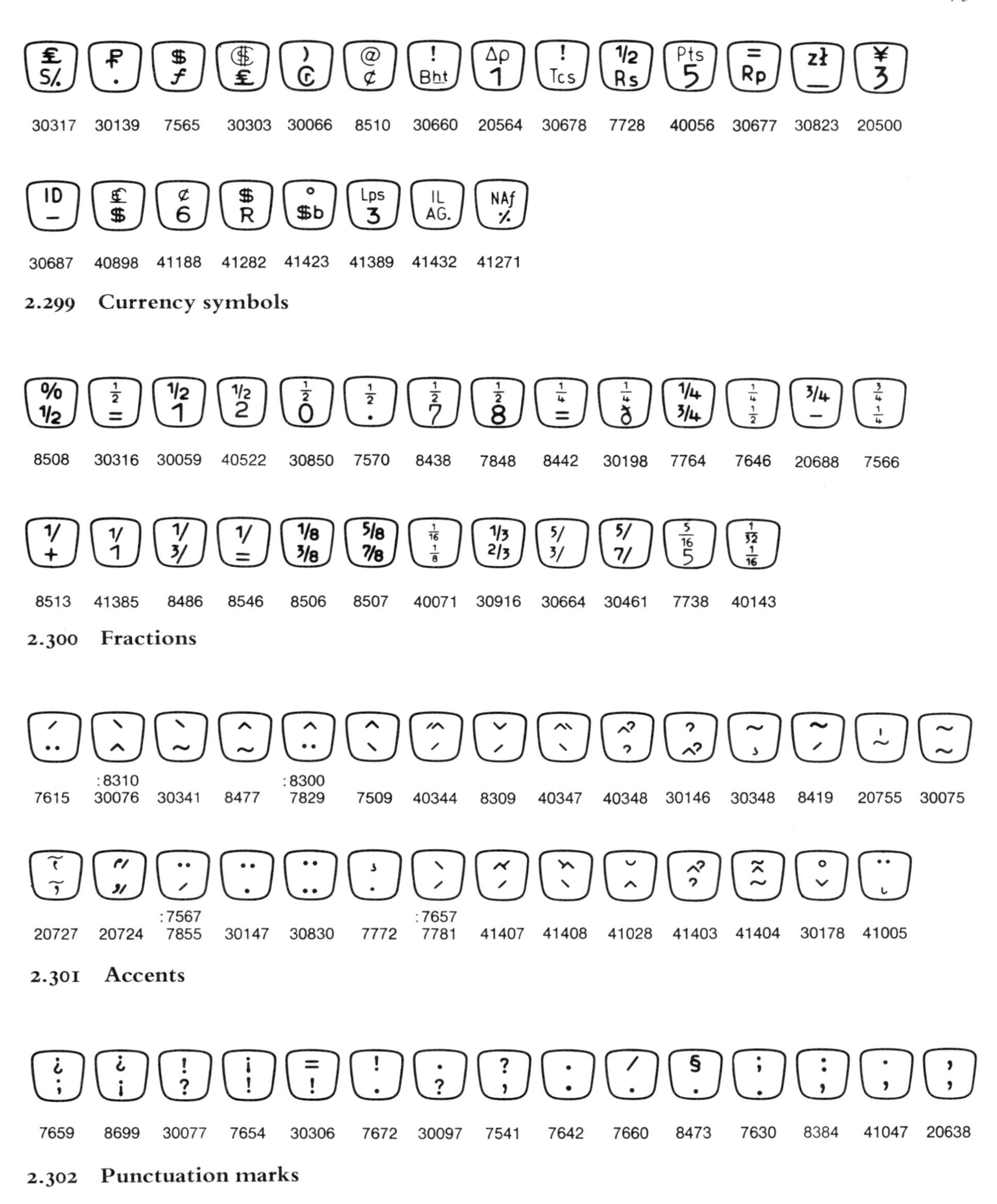

2.299 **Currency symbols**

2.300 **Fractions**

2.301 **Accents**

2.302 **Punctuation marks**

"TO SAVE TIME IS TO LENGTHEN LIFE"

Typewritten in 84 Languages

English–	To save time is to lengthen life.
French–	Gagner du temps, c'est prolonger la vie.
Portuguese–	Economisar tempo é alargar a vida.
Hungarian–	Takarékoskodj az idővel, meghosszabitod az életed.
Polish–	Kto czas oszczędza - przedłuża sobie życie.
Basque–	Demboraren irabaztia, biciaren luçatzia da.
Catalan–	Economizar tèmps es allargar la vida.
Provençal–	Temps gagna fa longo vido.
Breton–	Hastenn ar vuez ho c'honi amzer.
Irish–	Is Ionann Am-Coigilt agus Seagal-buanad.
Gaelic–	Faid saoghail is seadh do re chuir a b-feidhm.
Welsh–	Mae arbed amser yn estyn oes.
Manx–	Dy hauail traa te jannoo bea ny sleurey.
Flemish–	Tijd besparen is leven verlengen.
Frisian–	Tüd besparje is libjen verlenge.
Icelandic–	Að spara tíma er að lengja lifið.
Bohemian–	Úspora času jest prodloužením života.
Roumanian–	A economisi timp este a prelungi viaţa.
Slovenian–	Varčevanje s časom, je daljšanje življenja.
Slovak–	Ušporuvat čas je prodluhit života.
Esthonian–	Jõudsam tõõ on elu pidkendus.
Lettish–	Laiku taupot - pagarina dzivibu.
Lithuanian–	Užčėdyjimas laiko ilgina amžį.
Croatian–	Tko vrijeme štedi, taj produžuje život.
Spaniolish–	Economia di tiempu, alarga la vida.
Servian–	Тко вриjеме штеди, таj иродужуje живот.
Ruthenian–	Щѣнити часъ, то довше жите.
Bulgarian–	Спест ванкеявръме е увеличаване живота.
German–	Zeit sparen heisst das Leben verlängern.
Italian–	Risparmiando tempo prolungate la vita.
Latin–	Parcere tempori vitam longiorem facit.
Swedish–	Att vinna tid är att förlänga lifvet.
Danish–	At spare Tid er at forlænge Livet.
Norwegian–	At spare tid er at forlænge livet.
Finnish–	Aikaa voittaessa, elämä pidentyy.
Maltese–	Min jahdem fis, itaughal haghtu.
Albanian–	Kur ngi bier mot ron shum.
Romanch–	Spargner temp ais prolunger la vita.
Ido–	Sparar tempo esas longigar la vivo.
Greek (Ancient)–	Φείδεσθαι χρόνου ἐστὶ βίον μηκύνειν.
Greek (Modern)–	Ἡ οἰκονομία τοῦ χρόνου εἶναι παράτασις τῆς ζωῆς.
Esperanto–	Ŝpari tempon estas plilongigi la vivon.
Sioux–	Wicoran yuptecana kin he wiconi yuhanske.
Winnebago–	Wō shkännä lä kä lä kī cī gī shī, wankshik hō ī nä nī gī sā lētch nä nä.
Aztec–	Aquin àmo quixpoloa in cahuitl quihuellaquilia inemiliz.
	Ká taquick tiempo cu chokuactal á kimil.
Maya–	Ti pinagtiped iti añget paatidduguen ni biag.
Ilocano–	Magdaginot sa adlao, kay mao ang hataas ñga
Visayan–	kinabuhi.
Bicol–	Pag-imotan ang panahon pagpa-láwig nin buhay.
Pampango–	Ing pamagarimuhan king panaun makakaba king bie.
Pangasinan–	Say panagteper ed maong sa panahon so macasuldon ed pan bilay.
Tagalog–	Ang pag-aarimuhán sa panahón ay nakapagpapahaba ñg buhay.
Sizulu–	Lowo o gcina isikati sake u yena o nesikati eside ukusandisa emhlabeni.
Sesotho–	Ea sa senyeng linako tsa hae ke eena ea phelang halelele lefatseng.
Sixosa–	Ongaciti ixesha lake nguyena o nexesha elide ukulandisa emhlabeni.
Setshangaan–	A lavisaka shikati utomi wa yena u tayengeteleka muhlabeni.

2.303 A message typewritten in no fewer than eighty-four different languages

Russian—	Сберегая время удлиняемь жизнь.
Spanish—	Economizar tiempo es alargar la vida.
Dutch—	Tyd uitwinnen is zyn leven verlengen.
Taal—	Tijd te spaar maakt gebruik langer.
Quoc-Ngu—	Lợi ngày giờ, bằng sống lâu năm.
Hawaiian—	Malama pono ana‘ika manawa, He mea ia e hooloihi aku ai ike ola.
Maori—	E poto taima e ora roa.
Romanized Malay—	Me-niampurnakan waktu itu me-nambahi panjang umor.
Eskimo—	Uvdlunik aũngnertusârinek inûnertunarpok.
Hova—	Tsy mandany andro foana no manalava ny aina.
Arabic—	حفظ الوقت اطالة الحياة
Urdu—	هى وقت كا بچانا بڑھانا حيات كا
Malay—	ميمرناكن وقتو ايت منمباهى فنجڠ عمر
Persian—	وقت رامحافظت نمايى عمرت درازباشد
Sart—	وقتنكزني كمتسانكر عمرنكر ازايجاق
Tartar—	هركاه وقتى ملاحظه ايليمسن عمرون اوزون اولار
Turkish—	وقت قازانمق عمرى اوزاتمقدر
Sanskrit—	कालक्षेपवर्जनमायुष्प्रसरणम्
Hindi—	वख्त बचानेके मायने यह हय के जिंदगी बढाना।
Marawari—	समये का बचाणा उमर का बढाणा है
Magadhi—	काल का बचाना है आयु का बढाना।
Marathi—	वेळ वांचविणें म्हणजे आयुष्य वाढविणें आहे
Hebrew—	חשבון הקצר הוא ארוכת החיים.
Yiddish—	צייט שפארען הייסט דאס לעבען פערלענגערן.
Armenian—	[illegible]
Karen—	[illegible]
Burmese—	[illegible]
Japanese (Katakana)	[illegible]

TYPE AND TYPE STYLES

There are in the world today a number of factories devoted entirely to the manufacture of type and the main ones are Alfred Ransmayer & Albert Rodrian in Berlin, Setag SA in Switzerland, Novatype S.A. in Switzerland, and Tangens-Type GmbH in West Germany.

In addition, many typewriter manufacturers produce their own type and augment the range of type styles by also buying from various specialist firms.

Before World War II, most typewriters were equipped with the normal and familiar Pica typewriter type with ten spaces to the inch. The number of spaces to the inch is known as the 'pitch'.

During and after World War II, twelve spaces to the inch became more popular and type styles became much more sophisticated. Some examples of these are shown in the Figures 2.304–2.325. Each manufacturer continually adds to the range of type styles available. In addition, of course, there are the Cyrillic type styles (such as Russian), also Hebrew and Arabic where the carriage moves in the opposite direction, that is from right to left instead of the usual left to right.

2.304–2.325 Range of typefaces

PICA is one of the two main business type styles, excellent for stencil work because of its clarity.

A B C D E F G H I 1 2 3 4 5 6 7 8 9

PICA is a popular type style, easily readable, good for stencil work and copies.

A B C D E F G H I 1 2 3 4 5 6 7 8 9

ELITE is the other more usual business type style. Clear and neat for good appearance.

A B C D E F G H I 1 2 3 4 5 6 7 8 9

CONTINENTAL ELITE is slightly wider spaced, giving a clean and regular appearance.

A B C D E F G H I 1 2 3 4 5 6 7 8 9

CONGRESS PICA is the largest of the shaded types.
Gives a clean, impressive printed appearance.

A B C D E F G H I 1 2 3 4 5 6 7 8 9

CONGRESS MODERN PICA is similar in appearance,
but streamlined to give a modern printed effect.

A B C D E F G H I 1 2 3 4 5 6 7 8 9

CONGRESS MODERN ELITE is very popular, both for
private correspondence and business letters.

A B C D E F G H I 1 2 3 4 5 6 7 8 9

SMALL MICRO is even more condensed, with three
additional characters, making seventeen in all.

A B C D E F G H I 1 2 3 4 5 6 7 8 9

ITALIC PICA is quite distinctive, and legible.
Used for both business and private letters.

A B C D E F G H I 1 2 3 4 5 6 7 8 9

ITALIC ELITE is smaller and more compact,
and used rather more for private correspondence.

A B C D E F G H I 1 2 3 4 5 6 7 8 9

COPY ELITE is sharp, crisp and eventoned. It is suitable for correspondence and reproduction.

A B C D E F G H I 1 2 3 4 5 6 7 8 9

COPY PICA is sharp and crisp but the slightly wider spacing makes for easier reading.

A B C D E F G H I 1 2 3 4 5 6 7 8 9

Script is the perfect personal type style, for it resembles beautiful handwriting.

A B C D E F G H I 1 2 3 4 5 6 7 8 9

ITALIC ELITE BOLD is similar and smaller - quite distinctive and clean in appearance.

A B C D E F G H I 1 2 3 4 5 6 7 8 9

DOUBLE GOTHIC PICA IS MAINLY USED FOR BILLING AND ADDRESSING, BECAUSE OF ITS CLEAR CARBON COPIES.

A B C D E F G H I 1 2 3 4 5 6 7 8 9

DOUBLE GOTHIC ELITE GIVES MORE LETTERS TO THE INCH, AND IS EXCELLENT FOR STATISTICAL AND FINANCIAL WORK

A B C D E F G H I 1 2 3 4 5 6 7 8 9

CONTINENTAL CONGRESS ELITE with its slightly wider spacing gives an easily readable printed appearance.

A B C D E F G H I 1 2 3 4 5 6 7 8 9

CONGRESS ELITE is slightly more condensed,. but equally clear and impressive.

A B C D E F G H I 1 2 3 4 5 6 7 8 9

PIN-POINT PERFORATING TYPE IS THE PERFECT PROTECTION FOR CHEQUE-WRITING AND SIMILAR WORK.

A B C D E F G H I 1 2 3 4 5 6 7 8 9

TELEGRAPH TYPE IS BOLD AND CLEAR, AND MOST SUITABLE FOR TRANSMISSION OF MESSAGES.

A B C D E F G H I 1 2 3 4 5 6 7 8 9

DISPLAY is used for tickets, tags, labels and notices - any work demanding real attention.

A B C D E F G H I 1 2 3 4 5 6 7 8 9

CARE type is clear and neat and is especially suitable for statistical work and technical reports.

A B C D E F G H I 1 2 3 4 5 6 7 8 9

Generally the most suitable type for stencil cutting is Pica, or Elite; these are sharp and clear. Flat type faces are best for correspondence purposes where only one or two carbon copies are required.

In addition to the 10 or 12 pitch machines there are others such as five spaces to the inch (5 pitch) or seventeen spaces to the inch (17 pitch). Continental, British, and American spacings vary to some degree. In addition to these, there are also type faces where some letters occupy a greater space than others; this is proportional spacing. For instance, the letter 'm' takes up more spaces than the letter 'i'. This arrangement makes correspondence look as if it had been printed rather than typed.

All typewriter type is curved to match the curvature of the rubber cylinder. The type face is not flat as with printers' type.

The lower the price of a machine, the smaller is the range of type styles generally available. Some cheap portables are made with only one type style whereas some electric machines have twenty or thirty type styles from which to choose.

The obvious way to test a typewriter is to use each letter to see that it moves freely, but for some unknown reason people have always used strange sentences such as 'the quick brown fox jumps over the lazy dog's back', and countless families of foxes are still leaping lightly over legions of lazy dogs and will no doubt continue to do so for ever. The object of a test sentence is to use only twenty-six letters, and to the best of our knowledge, the only logical sentence so far conceived is the one found at the end of the next paragraph, in inverted commas.

There was a man by the name of James Quincey Vandz. He rather objected to the fact that his next-door neighbour kept a fox cub as a pet in his back garden. The cub grew into a rather large whelp and he said it smelt. One day Vandz threw a brick at it and hit the unfortunate animal. The owner reported Vandz to the authorities. When they asked him on what grounds he based his complaint, he replied, 'J. Q. Vandz struck my big fox whelp'.

There you have a genuine grammatical sentence using every letter of the alphabet only once!

One of the leading type and keyboard experts in the world is Mr. Tytell of U.S.A. His experience dates from 1937 when a departmental store required a Burmese typewriter for one of their customers. Tytell completed the order in five days working with type faces and the copy of a Burmese keyboard provided by a manufacturer. He set up in business and converted typewriters into machines for the particular requirements of individuals. His unusual creations include keyboards with planets for astrologers and a machine that writes in sixteen languages.

During World War II, one of his technicians, Max Burstein, a Russian, who had been a prisoner of war at Dachau survived only because the Nazis discovered he could convert captured Russian typewriters into German. After the war, he changed these same machines back to Russian!

As Tytell's business grew, he became more and more interested in the detailed study of typed documents. Typewritten document examination dates from the turn of the century and for several years, experts in this field were ridiculed or ignored, but by the 1920s, they were being listened to in court cases.

2.326 Definition of type style measurements

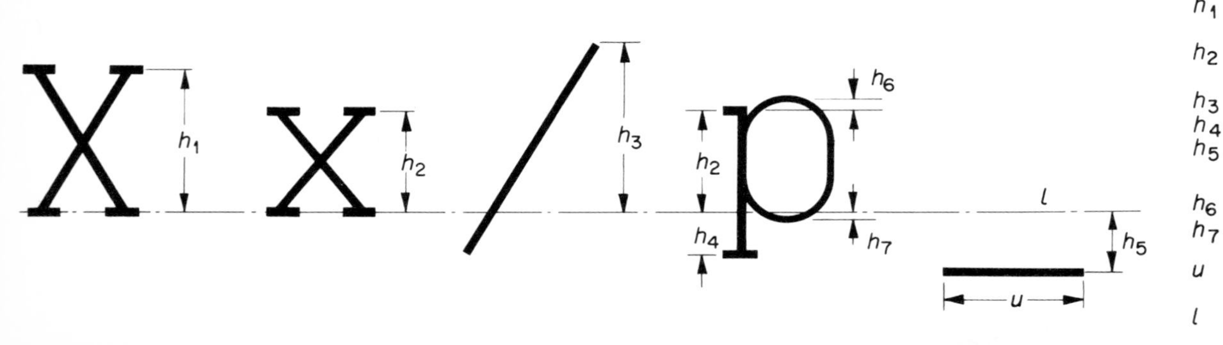

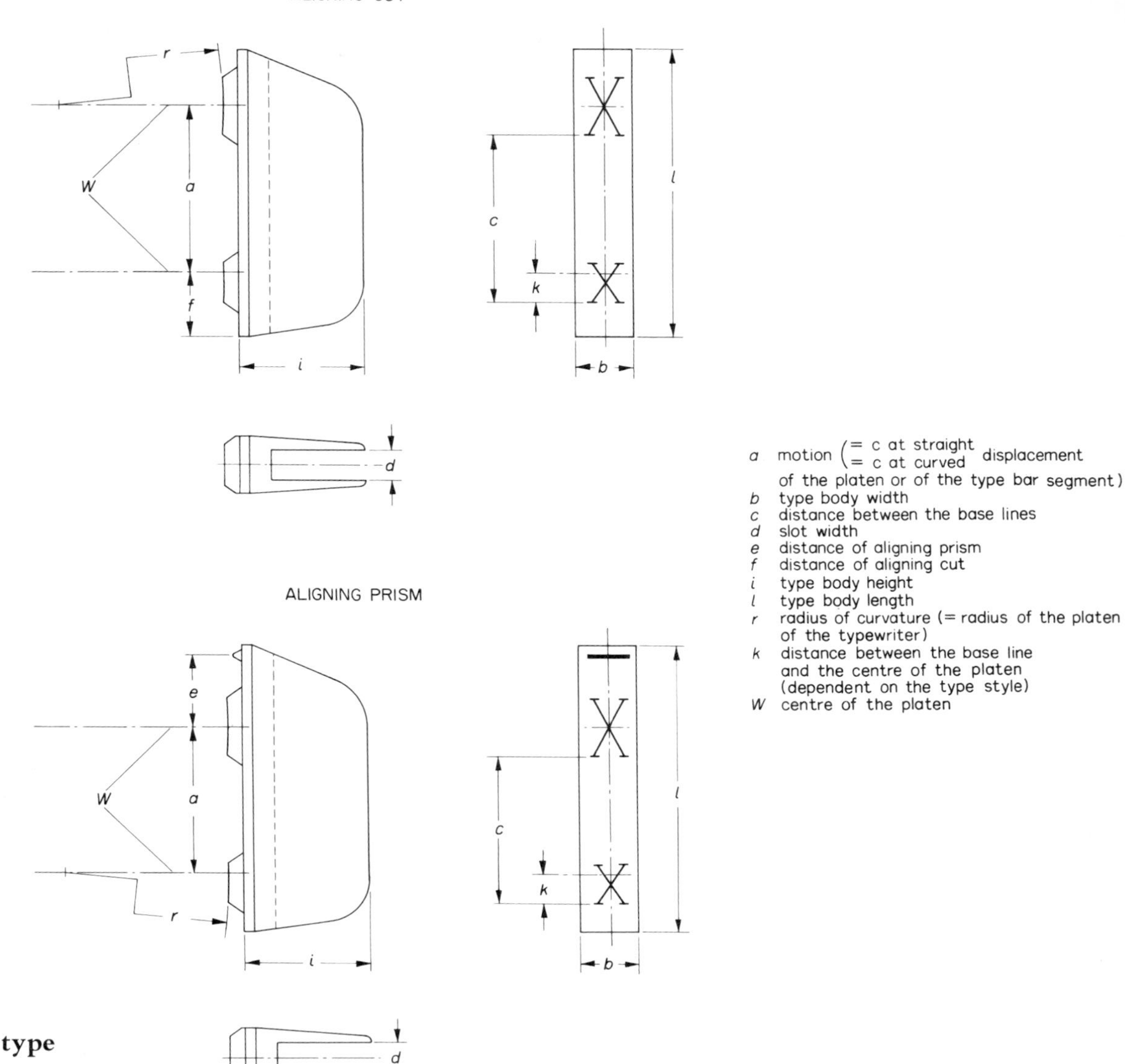

2.327 Definition of type body measurements

Tytell is convinced that typewriting is easier to trace than handwriting and warns people that they cannot hide their identities behind typewriters.

The criminal will often betray himself through some error or through some unconscious typing habit. Tytell's 'detective' work has been so successful that he has become known as the 'unofficial king of anonymous letter sleuths' and also as 'Mr. Typewriter'! There are similar people who work for Interpol in this particular field.

INKING RIBBONS AND CARBON RIBBONS AND THEIR VARIOUS ATTACHMENTS

Very early typewriters had great problems to overcome. Firstly, there was no clear understanding of the basic requirements. Most designers were obsessed with the idea of embossed writing, communication between blind people, and as stated earlier, using piano keyboards.

When it was realized that typewriters had a commercial future, a means of producing readable type was required, and this was achieved in the first place by the use of crude carbon

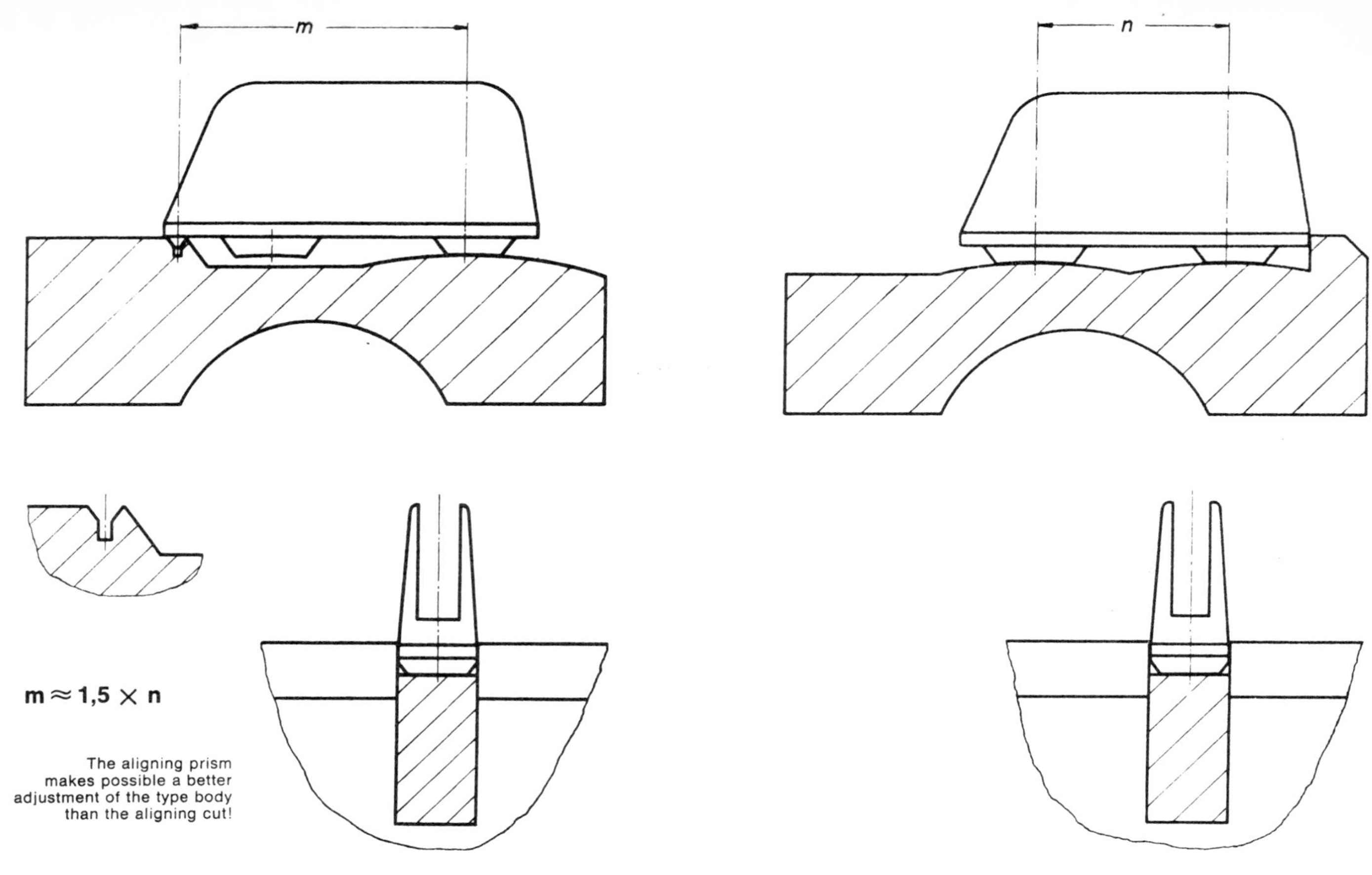
m
n
m ≈ 1,5 × n
The aligning prism
makes possible a better
adjustment of the type body
than the aligning cut!

2.328 Optical correction: type body with aligning prism

2.329 Optical correction: type body with aligning cut

sheets. Alexander Bain seems to have been the first person to produce a movable inked ribbon.

As the years went by, many curious devices were invented such as inked pads on which the type lay permanently—and incidentally corroded—or movable pads that passed over the various type faces. None of these were really satisfactory, and once the typewriter established itself, the ribbon in a crude form became, with few exceptions, the standard method of transferring the ink to the paper. It also had the advantage of allowing carbon sheets to be placed between the top and other copies, thus giving a readable impression and enabling the writer to retain a useful copy for record purposes.

The production of ribbons, both single- and two-colour, became a thriving business. Typewriters once purchased became an almost permanent fixture in an office, but carbons and ribbons were needed continuously and, in many cases, those who produced them became more prosperous than the designers and manufacturers of machines.

Manufacturers of early machines sought to produce an individual ribbon spool to compel those who had purchased machines to continue to buy supplies. On certain models, the left spool differed from the right spool.

Early up-strike machines had wide ribbons, but by about 1900 most machines were using either $\frac{1}{2}$ inch or 13 mm widths.

Today there is still a wide variety of spools, but German manufacturers standardized theirs very early and, consequently, all their ribbon spools, whether left or right, were interchangeable between any machines. They are known as the 13 mm DIN or 'NORM' spool.

The carbon ribbon also made its appearance with varying degrees of success. This consisted of a large 8 mm spool of carbon paper, which only went through the machine once but gave a

a

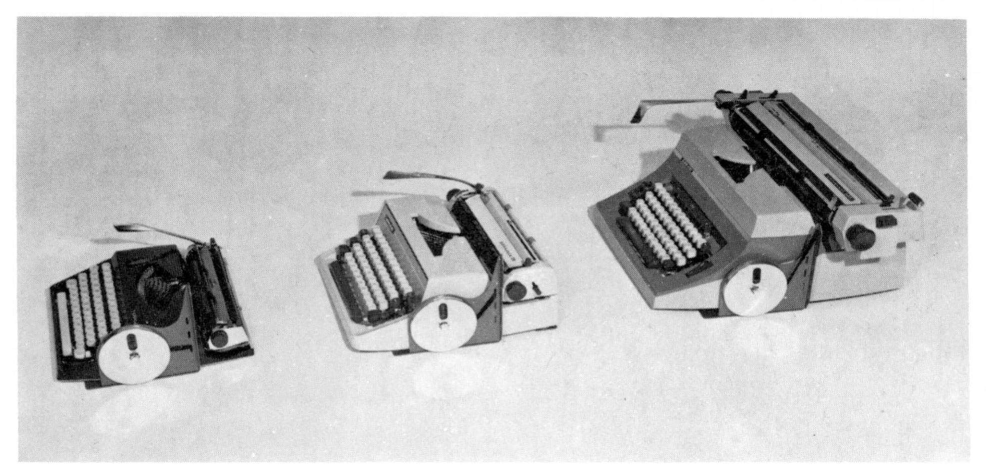

b

Ordinary typing on Ordinary Ribbon

Carbon Ribbon typing, with attachment.

c

2.330a SB Universal Carbon Ribbon Attachment
b Attachment fitted to machines
c Effect on typing

sharper, clear impression. At first great trouble was encountered. Either the ribbon drive was too slow and the typing irregular, or the ribbon drive was too fast and the carbon ribbon, being made of paper, was continually breaking. In addition, it was only possible to write in one colour. Therefore, credit notes and various other items could not be prepared on such a machine.

About 1960, however, new plastics were produced. These were not only clearer and thinner and more efficient, but had the advantage of not breaking or tearing so easily. They were first introduced on electric machines and later became associated with electric typing. In this respect the electric typewriter had an advantage of evenness.

About 1965 some manufacturers devised a means of equipping electric typewriters with both a two-colour ribbon and a carbon ribbon so that the typist may switch from one to the other. This was a great step forward in design and is now common to most top quality electric typewriters. Many manufacturers, however, persisted in designing carbon ribbon attachments that would operate on manual machines successfully and considerable numbers of these were sold. Furthermore, a number of individuals, both in England and in Switzerland, endeavoured to produce a carbon ribbon attachment which could be fitted to almost any machine. This would enable a manual typewriter to convey the impression of 'electric typing' which the public had come to associate with the use of a carbon ribbon.

The most difficult problems to overcome were:

1. The fitting of the carbon ribbon attachment to the side of the machine.
2. The speed of the ribbon drive.

The most successful of these is the one illustrated at Figure 2.330, which is manufactured in England and is known as 'The Universal Carbon Ribbon Attachment'; it is free standing and operates on most modern machines where the ribbon drive is fast enough.

The strangest thing of all, of course, is that a full circle has now been completed. Typing with carbon paper preceded the ribbon, but most people now prefer carbon ribbon typing which gives greater clarity, as the quality of the product is far superior.

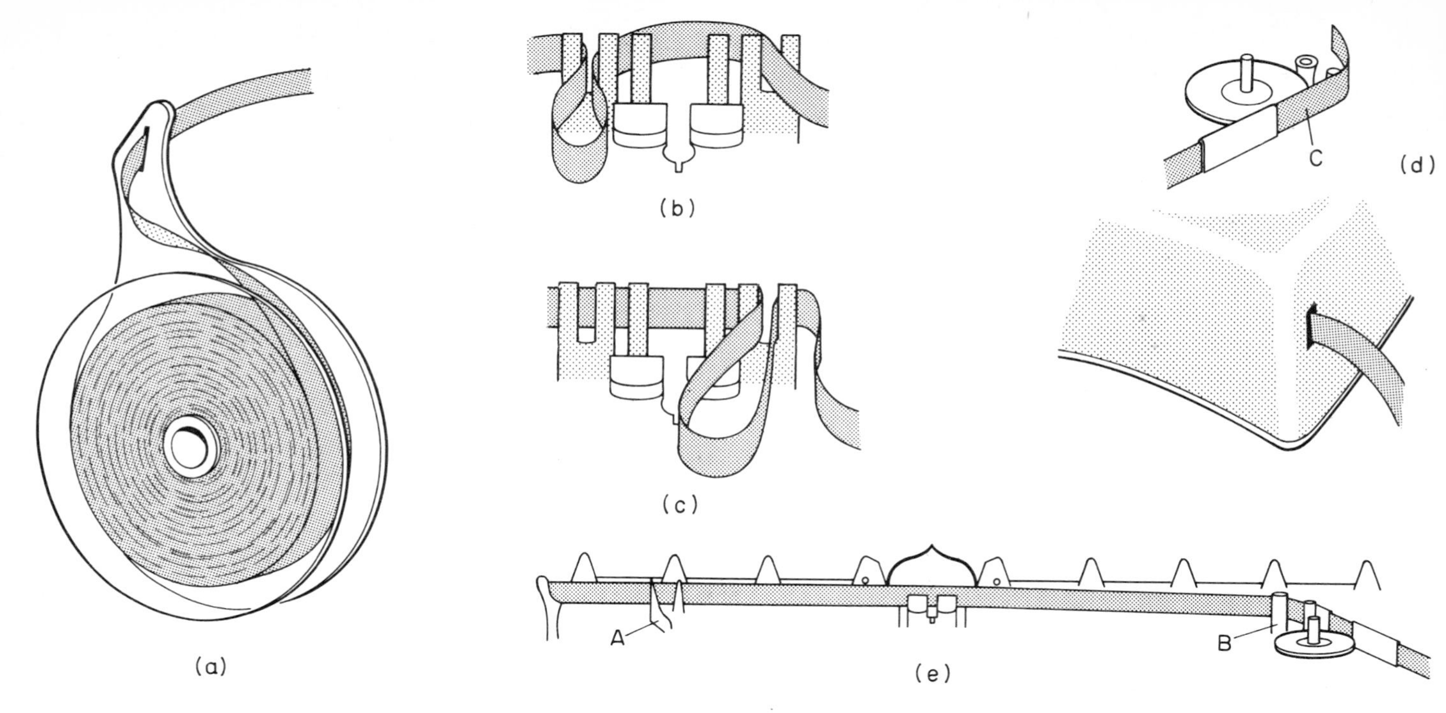

2.331 Carbon ribbon attachment patented in 1955 by W. Rolf Pedersen A/S of Copenhagen, Denmark

There is very little difference in cost between using a good modern plastic carbon ribbon and a good quality silk or nylon ribbon. Cheap cotton ribbons give a poor impression with a blurred outline and wear out quickly. They are probably the most expensive of all in the long run.

There are many other ideas for inking or producing the readable final impression, but at this stage, they are still only experimental. No doubt in the next few years, some revolutionary process will emerge as it has already done with many items in common use. Whereas once the transition was slow, as from the horse to the internal combustion engine, today it is rapid and dramatic as was the change from electric mechanical calculators to electronic calculators.

Franz Buttner AG of Zurich, Switzerland invented a carbon ribbon attachment which was patented on 15 September 1966 but this has not been produced.

W. Rolf Pedersen A/S of Copenhagen, Denmark, patented a carbon ribbon attachment in 1955. This is illustrated in Figure 2.331. It was called the carbograph and had moderate success but has been discontinued.

Imperial also produced a carbon ribbon attachment which was fitted to the Model 55 in about 1954. Other manufacturers made similar attachments but later abandoned them because electric machines were being produced with built-in carbon ribbons and such an attachment on a manual machine interfered with the sale of electrics.

ANNUAL PRODUCTION OF TYPEWRITERS

The table No. 1 shows the total apparent consumption of typewriters in the U.K. One of the most interesting aspects of these figures is that in 1972, for the first time, the number of electric typewriters exceeded the number of standard manuals sold. It is also interesting to see that the sales of portable typewriters continue to increase. Eventually, of course, electric and electronic machines may well take over the field completely.

Although most typewriters seem to be virtually indestructible, and in spite of the introduction of photo-copying machines which almost all commercial firms are now widely using in place of copy typists, these figures show that the demand for typewriters increases every year.

It is believed that the annual production of typewriters in the world is between 8 and 9 million. This is only an approximate figure as most manufacturers are, for some reason or other, reluctant to give accurate details of their own production.

Table 1

TOTAL APPARENT CONSUMPTION OF TYPEWRITERS IN U.K.

Year	*Standard*	*Portable*	*Electric*
1952	58,500	43,505	450
1953	56,500	58,432	1,000
1954	80,000	67,253	1,390
1955	96,250	74,234	2,370
1956	93,800	79,355	5,080
1957	97,200	100,365	5,385
1958	88,500	109,012	4,744
1959	98,400	141,691	7,435
1960	125,500	141,573	12,700
1961	119,000	141,923	12,700
1962	103,456	139,738	15,505
1963	101,380	180,810	16,402
1964	116,848	206,790	27,930
1965	111,450	220,640	29,601
1966	97,000	220,000	39,000
1967	90,000	220,000	44,000
1968	90,000	230,000	49,000
1969	89,000	240,000	51,000
1970	86,000	250,000	54,000
1971	82,000	280,000	70,000
1972	(*est.*) 74,000	300,000	80,000

THE AGE AND EFFICIENCY OF A TYPEWRITER

Very few typewriters ever appear to be thrown away and consigned to the flames or the dust-bin. Some find their way to the auction sale and come under the hammer, and many are rebuilt—some of them on more than one occasion.

Typewriters can last for fifty or sixty years or even more and, it may perhaps be said, seem almost indestructible. One American typewriter company managed to lay hands on the first machine they ever made, and as a sales promotion operation, dressed a typist in a crinoline and placed her in the corner of their main showroom, typing envelopes on the machine, with a notice which read 'Please do not ask us how long our typewriters last. This lady is using the first typewriter we ever made and we are still trying to find out!!!'

Too much importance is often attached to the age of a typewriter. It is true, of course, that as the years went by certain manufacturers made better models than their competitors, but the main groups of machines were well constructed and varied very little in their performance, producing good, trouble-free work for many years.

It must be emphasized that the quality of work produced is little influenced by the age of the machine. Type styles have become more sophisticated, but it must be obvious that a typewriter can be twenty or even fifty years old and still produce excellent work if it has only had limited use, whereas a machine that is only two or three years old can be virtually worn out if it has been in constant use in a busy office from 9 a.m. to 5.30 p.m. five days a week.

The introduction of the plastic keyboard instead of the key card and key ring is an advantage for ladies with long finger nails, and the elegant décor and colouring of certain machines may blend with the modern office; yet, although a machine with the well established 'QWERTY' keyboard, two-coloured ribbon and backspacer may not enhance the décor of a modern office, it could easily be more reliable and efficient than a machine of streamlined elegance which has been hammered on daily for years by a heavy-handed typist. Obviously the most important consideration is the quality of work the typewriter is capable of producing and the degree of its reliability.

It is also worth pointing out that a good reconditioned typewriter is generally a better proposition than a used car. There are far more second-hand typewriters being efficiently used than there are second-hand motor cars of the same age. No four-bank typewriter should be condemned solely because of its age. In this connection it is worth pointing out that machines should be kept in warm, dry offices and should be regularly maintained for they suffer more from neglect than from use.

TYPEWRITERS IN USE TODAY

Apart from a few experimental models of varying degrees of efficiency there were no typewriters in use in 1873. In that year, however, Remington set the wheels of the typewriter industry in motion with the first commercially produced standard office machine. The company manufactured 1,000 machines in the first year.

Other manufacturers followed but the total output of typewriters in the world increased quite slowly, and ten years after the introduction of the first Remington, there were still relatively few machines in use throughout the world.

However, the swing towards the typewriter began to gather momentum, and at the beginning of the twentieth century the demand for typewriters had appreciably increased. More than thirty manufacturers were producing typewriters in America alone.

In all, over 400 different makes of machines have so far been produced, of which some 140 have been made in America. Europe is a comfortable second in the league table of production, with Germany, followed by the United Kingdom playing leading roles, while Japan has vastly increased its output of typewriters since World War II.

Typewriters are also manufactured in India and China. All these countries are in the Northern Hemisphere.

No typewriters are actually manufactured south of the equator, although there are plants in South America for assembling Olympia typewriters and a similar plant in Johannesburg, South Africa, owned by Olivetti.

It was the front-strike machine with the 'QWERTY' keyboard and four rows of keys that finally dominated the market, and all those manufacturers who were not prepared to conform were ultimately squeezed out by vicious competition, or simply defeated by lack of public demand.

Nowadays there are approximately 120 different keyboards based on the 'QWERTY' arrangement—but modified in various ways to suit different languages. Similarly other modifications can be made to solve problems set by foreign languages—as in the case of Hebrew and Arabic. For these languages typewriters are made with a carriage moving from left to right.

Some typewriter companies lasted many years and their names became houseworld words such as 'Oliver', 'Yost', 'Barlock', 'Smith Premier', 'Densmore', 'Hammond', 'Blickensderfer', 'Noiseless', 'Empire', etc., but gradually, under the pressure of fierce competition, they disappeared. Others lost their identity as they were swallowed up by larger organizations, such as Remington, who claimed that by 1954 they had produced over 15,000,000 typewriters. Many firms struggled on for years hoping against hope, but finally gave up the unequal fight and went bankrupt or into voluntary liquidation.

Some really never got off the ground at all, the most spectacular of these being the 'Conqueror', an English venture. This was a copy of the German 'Stoewer' financed after World War II by Lord Lascelles, son-in-law of King George V. It is said that when the King heard about it, he forbade any members of the royal family to engage in commerce or industry of any kind, and so the project was abandoned. One machine only was produced, but as the whole assembly line for full-scale production had been prepared, it is estimated that the typewriter must have cost well over £1,000,000. It conquered nothing. What was conceived as a colossus proved to be a helpless pygmy!

Today there are about twenty-five typewriter manufacturers in the world; and together they produce annually an estimated 8–9 million typewriters, electric, standard, and portable. The bulk of these are produced by the following seven large groups which are vertically integrated combines encompassing the world:

Brother manufacture in Japan and sell, under their own name, compact electric machines and a large range of portables throughout the world. They also manufacture a large range of appliances such as sewing-machines, washing machines, dishwashers, electric motors, adding machines, and electronic calculators. Some of their typewriters are marketed under the name of 'Remington' in various parts of the world. Latest available figures show that even though their typewriter production is approaching the million mark, it represents only 12½ per cent of their industrial production.

International Business Machines Corporation known as *IBM* who produce in the U.S.A., Germany, France, Holland, and England. They make neither portable nor manual machines, but concentrate on standard electric typewriters which they sell directly to the public through their own sales organization. They also manufacture computers and a multiplicity of space-age technical equipment, much of which is based upon the typewriter.

Litton Industries own and control 'Adler' and 'Triumph' in Germany and Holland, 'Royal' in America, and 'Imperial' in England. They take typewriters produced by the Silver Reed Knitting Machine Company in Japan, as well as portable typewriters manufactured in Portugal. These machines are sold under the four brand names in different parts of the world. Typewriter sales, though large, are a relatively small part of Litton activities.

Olivetti/Underwood Corporation manufacture and assemble electric, manual, and portable typewriters, together with a multiplicity of adding machines and electronic office equipment in Italy, Spain, England, and various other countries. They sell their goods under both brand names throughout the world.

Olympia manufacture in Germany, and have various assembly bases in South America, Yugoslavia, and Ireland. They also manufacture adding machines and sophisticated electronic machines.

Remington Rand Ltd. manufacture in the United Kingdom, America, Italy, Holland, and Germany. They also sell 'Brother' machines which are made in Japan. Remington Rand make many other sophisticated products.

Smith-Corona manufacture electric and compact portables in America and portables in England, but no longer produce standard manual typewriters. They market many other products as well throughout the world.

The smaller groups include such companies as: Hermes in Switzerland, Facit in Sweden,

Messa in Portugal, Kovo in Czechoslovakia, and the combined East German factories producing such machines as 'Erika' and 'Optima'. 'Lucznik' is produced in Poland, 'Godrej' in India, and the 'Nippo' in Japan. A considerable number of manufacturers sell their machines under different brand names, which vary from country to country and sometimes from year to year.

The currency regulations of many countries impose restrictions on the importation of typewriters, but generally speaking, most of the well-known makes are available, under one name or another, in most parts of the world, and often in the most surprising places!

There is very little detailed information concerning the production of typewriters in China or Russia, but we do know the Russians make the 'Moskva' portable, the 'Progress' and 'Baskiria' office machines, and the 'P.E.K. 45' office electric machine. They also import typewriters from East and West Germany.

The Yugoslav metal industry concern UNIS, near Sarajevo, manufacture under licence for Erika, the Biser, a semi-standard machine, and also under licence for Olympia the SF Traveller and Traveller de Luxe.

Typewriter production depends upon a vast number of subsidiary industries that make certain parts as sub-contractors. This can best be illustrated by the story of the Continental 'Silenta'. This was originally made in a factory in Chemnitz which, it is reported, was transported en bloc together with many technicians to Russia immediately after the war. The machine was far ahead of its time, and probably one of the best ever manufactured, but only a few were made and they appeared after the war in various parts of Europe. With the best will in the world, it was impossible to manufacture this machine in Russia satisfactorily. The know-how of the sub-contractors, who had been mostly in West Germany, producing typewriter parts such as type guides, escapement wheels, etc. had gained invaluable experience over the years, but they were no longer available. Many had been killed and the technical know-how and location of the source of their materials was a closely guarded secret which it was impossible to unravel. It would appear that the project was finally abandoned.

A similar fate befell the Continental portable assembled in Belgium immediately after the war. All the machines looked the same but they did not work the same—or if they did—not for very long.

There are approximately forty-two operations in the manufacture of each typebar alone. From this it can be deduced that typewriter production is a manufacturing nightmare. It is quite true to say that today, with any given quantity of capital, manufacturing capacity and skill, almost anything is more profitable to manufacture than a typewriter.

It is significant that the lion's share of the market is controlled by those who regard the typewriter not as an end in itself, but as a means to an end—something through which a more sophisticated machine can be developed.

POSSIBLE ORIGINS OF NUMERALS

The shape of the letters of the alphabet appear to be illogical in that they have no relation to sound. The only exception is the letter 'o' which is roughly the shape of the lips as the sound is made. Conversely numerals appear logical. At first Roman numerals were used, i.e. I, II, III, IV, V, VI, VII, VIII, IX, X, etc.

This was obviously difficult for calculations. The figures now used are Arabic in origin and are completely logical:

ZERO (0) is represented by a circle and therefore has no strokes at all.

A further equally logical and possible explanation is the number of angles used, as illustrated opposite:

Again, zero (0) has no angles. Strangely enough the new peripheral type on the bottom of cheques and credit cards follows the pattern as it is a language the machine understands.

1	One stroke =		1	One angle =
2	Two strokes =		2	Two angles =
3	Three strokes =		3	Three angles =
4	Four strokes =		4	Four angles =
5	Five strokes =		5	Five angles =
6	Six strokes =		6	Six angles =
7	Seven strokes =		7	Seven angles =
8	Eight strokes =		8	Eight angles =
9	Nine strokes =		9	Nine angles =

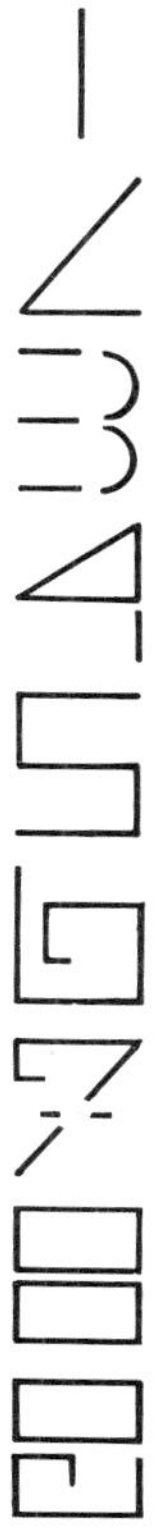

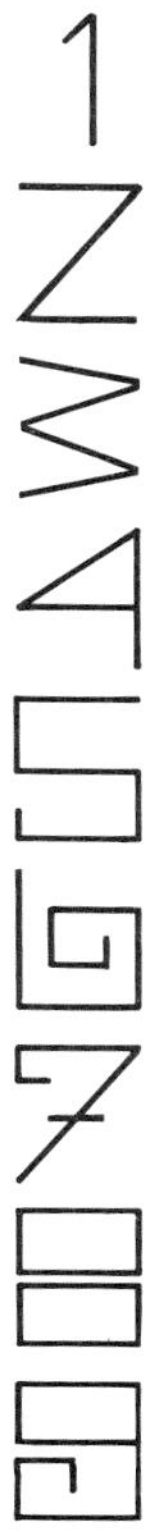

SOME ODD AND INTERESTING FACTS ABOUT TYPEWRITERS

- There are thirty-six typewriters on the QE2 including six for the use of passengers.
- An American aircraft carrier has approximately ninety typewriters.
- Military aircraft in both World Wars carried typewriters as it is easier to type in turbulent conditions than it is to write.
- Service of a secretary with a typewriter is available on certain German trains. British trains had a similar service but the last train to leave with a typewriter and a typist for the use of passengers was in 1914.
- The variation of makes, models, carriage lengths, and type styles available today, combined with approximately 147 different keyboards, gives a mathematical possibility beyond belief.
- An electric typewriter has approximately 3,700 parts, a manual 2,000, and a portable 1,000.

× CHAPTER 3

History of Typewriter Manufacturers Part I

3.1 Heinrich Kleyer (1854–1932)

ADLER: GERMANY

The Adler typewriter was developed by the Adlerwerke, previously known as Heinrich Kleyer. Kleyer had been employed by a machinery-importing organization in Hamburg, but in 1879 he resigned his position and crossed the Atlantic. After finding employment in several factories in America, he returned to Germany in 1880 and, benefiting from the knowledge and experience he had gained, he set up his own business, manufacturing bicycles in Frankfurt am Main.

He was soon interesting himself in typewriter production and in 1896, he acquired the patent of Kidder's 'Empire' machine. By this time Kleyer had turned his works into a 'limited liability company' and the factory became known as the Adlerwerke.

The first typewriters he produced in 1898 were based on Kidder's invention of the horizontal push, or forward thrust action, and bore the name 'Empire' after the American, *not* the Canadian machine of that name, but he was dissatisfied with its performance, and within two years had effected substantial improvements. The name 'Empire' was then dropped and 'Adler' substituted in its place, and every machine had to pass a severe test for quality.

In their early days Adler typewriters were distributed and sold through bicycle dealers. This frequently happened with other typewriters in other countries, but in the case of Adler the connection could be clearly seen as their trade mark included not only the *Adler* (i.e. 'Eagle') but also a bicycle wheel. This is now incorporated in a different position in the centre.

The company prospered: by 1909 some 50,000 machines had left the Adler factory, and

Adler Manual Standard Typewriters

3.2a First model of 1898

3.2b Model 7 of 1900

c

d

3.2 Adler Manual Standard Typewriters (contd.)
- c Model 8 of 1902
- d Model 21 of 1913. First machine produced with a four-bank keyboard by Adler on the Kidder principle

a b c

d e f

3.3 Adler Manual Standard Typewriters
- a Model 7 of 1900
- b Model 25 of 1925
- c Model 31 of 1931
- d Model of 1938 (Serial No. 570, 160)
- e Universal of 1951
- f Manual of 1963

contd. overleaf

Adler Manual Standard Typewriters (contd.)

3.3g Model 390 of 1963 onwards

3.3h Special of 1951

3.3i Special 13 in. of 1972

3.4 Adler Portable Typewriters (see opposite)

a

b

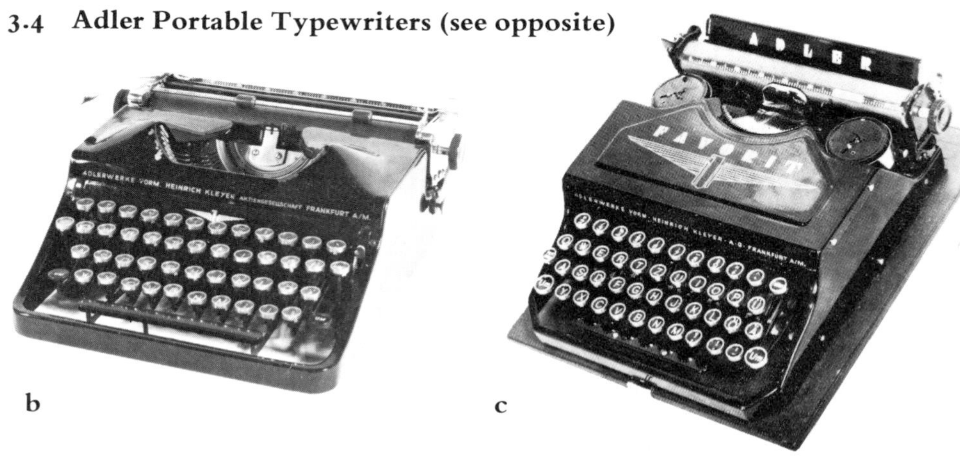

c

d

e

f

g

h

i

3·4 **Adler Portable Typewriters**

- a **Model 1 of 1912**
- b **Model 32 of 1932**
- c **Favorit Model 1 of 1935**
- d **Model of 1935–6**
- e **Gossen Tippa (later became Adler) of about 1938–9**
- f **Primus/Junior 3 of 1959**
- g **Gabriele 25 of 1968**
- h **Gabriele 35 of 1968**
- i **Gabriele 10 of 1970**
- j **Tippa S of 1970**
- k **Tippa 1 of 1972**

j

k

the number had doubled by 1913. The price then of an Adler was 300 marks—about fifteen times the weekly wage of the ordinary working man in Germany.

In 1913 a smaller machine, the 'Klein-Adler' was designed and produced. This was a portable version of the Model 7 standard machine, and was supplied with a case. It was sold in many countries, under different names: in Italy as the Adler 'Piccola', in Spain as the 'Adlerita', and in France as the 'Adlerette'.

During World War I the Adler Works were compelled to switch their production to articles more vital for the prosecution of the war, and among other things they manufactured cable-winding motors, lorries, ambulances, and other special vehicles.

a b c d e

3·5 **Adler Electric Typewriters**

- a **L & S Electric of 1957**
- b **Model E20 of 1961**
- c **Model 21C/41C of 1964**
- d **Model 21D/21F of 1972**
- e **Gabriele 5000 (compact) of 1972**

Table 2

TYPEWRITER MODELS PRODUCED BY TRIUMPH ADLER VERTRIEB

		Adler Standard Manual		Adler Electric		Adler Portable	
		Universal		**L. & S.**		**Privat**	
24,000	1906	800,001	1950	6,000,046	1957	3,000,001 to	1953
28,000	1907	825,590	1951	6,001,001	1958	3,008,000	
40,000	1908	872,701	1952	6,002,101	1959	3,008,001 to	1954
50,000	1909	920,000	1953	6,005,221	1960	3,018,000	
62,000	1910	962,001	1954	6,012,441	1961	3,018,001 to	1955
75,000	1911	1,010,501	1955	6,119,056	1962	3,027,300	
90,000	1912	1,065,501	1956	6,125,255	1963	3,027,301 to	1956
102,000	1913	1,125,001	1957	6,127,541 to	1964	3,035,150	
112,000	1914	1,177,001	1958	6,128,308		3,035,151 and up	1957
115,000	1915	1,224,001	1959				
120,000	1916	1,275,001	1960	**'S'**		**Primus/Junior 3**	
124,000	1917	1,329,701	1961	6,206,000	1960	1,630,000	1959
155,000	1918	1,394,891 to	1962	6,212,001	1961	1,636,901	1960
170,000	1919	1,431,600		6,221,001	1962	1,644,131 to	1961
190,000	1920	8,017,101	1963	6,225,001	1963	1,649,900	
212,000	1921	8,074,901	1964	6,227,001 to	1964	3,141,120 to	1961
229,000	1922	8,160,107	1965	6,227,852		3,143,000	
280,000	1923	8,266,998	1966			3,144,000 to	1962
280,000	1924	8,371,564	1967	**E.20**		3,147,000	
300,000	1925	8,435,833	1968	1,735,001 to	1961	3,180,000	1963
330,000	1926	8,547,920	1969	1,736,730 and		3,319,601	1964
350,000	1927	9,030,826	1970	7,100,000	1961	3,522,870 to	1965
360,000	1928	9,090,009	1971	7,101,586 to	1962	3,546,000	
400,000	1929			7,103,421			
410,000	1930	**Special**				**Favorit**	
420,000	1931	2,000,001	1951	**E.21**		1,654,400	1959
430,000	1932	2,001,751	1952	7,110,000	1962	1,655,251	1960
440,000	1933	2,020,002	1953	7,111,734 to	1963	1,659,641 to	1961
450,000	1934	2,036,001	1954	7,119,999		1,661,000	
465,000	1935	2,070,001	1955	7,140,000 to	1964	3,170,000 to	1962
490,000	1936	2,106,401	1956	7,144,999		3,176,000	
519,000	1937	2,134,501	1957	7,145,000	1965		
544,000	1938	2,153,001	1958	7,407,670	1966	**Junior E**	
582,000	1939	2,180,001	1959	7,469,113	1967	1,604,000 to	1959
617,000	1940	2,190,001	1960	7,517,923	1968	1,608,000	
625,000	1941	2,234,461	1961	7,593,046	1969	1,608,601 to	1960
720,000	1948	2,261,421	1962	7,676,056	1970	1,615,275	
825,000	1950	2,292,801	1963	7,827,299	1971	1,617,000 to	1961
900,000	1952	2,320,801	1964			2,123,000	
		2,346,678	1965	**E.41**		3,121,000 to	1962
		2,377,672	1966	7,130,000	1963	3,192,000	

cont. page 97 *cont. page 97* *cont. page 97*

Table 2 cont.

Adler Standard Manual	
2,412,542	1967
2,423,559	1968
2,456,338	1969
2,484,305	1970
2,494,501	1971

Adler Electric	
7,130,818 to 7,131,999	1964
E.21C/41C	
7,301,000	1964
7,344,175	1965
7,410,315	1966
7,469,086	1967
7,532,736	1968
7,604,331	1969
7,698,675	1970
7,870,530	1971

Adler Portable	
3,192,001 to 3,322,600	1963
3,322,601 to 3,522,869	1964
3,522,870 to 3,660,000	1965
Tippa I & S	
4,170,000	1959
4,195,001	1960
4,210,451	1961
4,277,000	1962
4,332,001	1963
4,395,101	1964
4,526,458	1965
4,592,513	1966
4,701,754	1967
4,835,621	1968
5,008,015	1969
5,125,873	1970
5,176,672	1971
Junior 10/20/30	
3,611,000 upwards	1965
(Name 'Junior' replaced by 'Gabriele' in September 1966)	

The production of typewriters resumed after the war, and by 1922 the annual production had been stepped up to 32,000—a record which was not excelled for sixteen years. This figure had been reached just after disastrous inflation had hit Germany—a time when people preferred to own goods rather than money. Prices rose to an astronomically high figure and went on increasing, and deliveries of typewriters were made on condition that goods were paid for on the day they were delivered.

The trade depression in the early 1930s spread from America to Europe. Germany too was seriously affected. The number of typewriter orders received by Adler fell from 22,000 to 5,000 annually. Several million people were unemployed in Germany at this time.

In 1931 the Adler 'Standard'—Model 31—was produced with segment shift. The depression was over, and the fortunes of the company changed for the better again, and in 1939, 37,000 machines were manufactured. War once again interfered with the output of typewriters, and the productive capacity of Adler was shared between typewriters and other instruments, such as teleprinters and telephonic components required by the government. Worse was to follow. On 20 March 1944 an air raid destroyed the whole assembly shop. By 8 May 1945 about four-fifths of the factory was destroyed, with the result that production of Adler's bicycles, motor-car components, and typewriters could be continued only on a much reduced scale.

After the war, in 1946, a new factory was built and typewriter production resumed. Progress was rapid, and in quick succession Adler produced the 'Standard', 'Universal', 'Special' and in 1952 the Adler 'Privat', a small portable typewriter. In 1954 the first Adler electric typewriter was shown at the Hanover Fair, and by this time, the company had made up the losses of the war years, exporting to more than eighty countries and re-establishing connections with Eastern Europe.

In 1956 the Company acquired the production rights of the Adler 'Tippa'. This was one of the most popular small portable typewriters with a plastic outer casing. It won a gold medal in Milan in 1960 and was classified in many areas of the world as the best flat portable.

In November 1957 Max Grundig, famous for radio and tape recorders, took over the Triumph factory in Nuremberg, and Adler joined forces with Triumph under his direction and control.

In 1966 Triumph acquired most of the working capital of Adler, and by the end of 1968, it had 82 per cent of the shares, by which time both Adler and Triumph were being sold in various parts of the world. They were identical machines under two different names.

On 10 January 1969 the share capital of both Triumph and Adler was acquired by Litton Industries.

Adler no longer produce bicycles, motor cars, or aircraft; they are active only in the office equipment field, with ever increasing success, enjoying a world-wide reputation for the production of typewriters, electronic book-keeping machines, electronic calculators, and sophisticated office machinery, as well as a complete range of portable and electric portable typewriters.

Factories are producing Adler typewriters in Germany and in Holland under the control of Litton Industries. T.A.V. (Triumph Adler Vertrieb) Distribution, as they are now known, have a world-wide representation and are producing and selling machines internationally in ever increasing quantities.

Some Adler/Triumph electric typewriters are sold in America as 'Royal' and in England as 'Adler', 'Triumph', and 'Imperial' under the direction of Litton Industries.

BROTHER TYPEWRITERS: JAPAN

3.6 Brother Electric Typewriter Model 3015 of about 1968

3.7 Trade Mark of Brother Industries Ltd. ('m' of MACHINE is the basis for the design of the initial Chinese letter of the establisher Yasui's name)

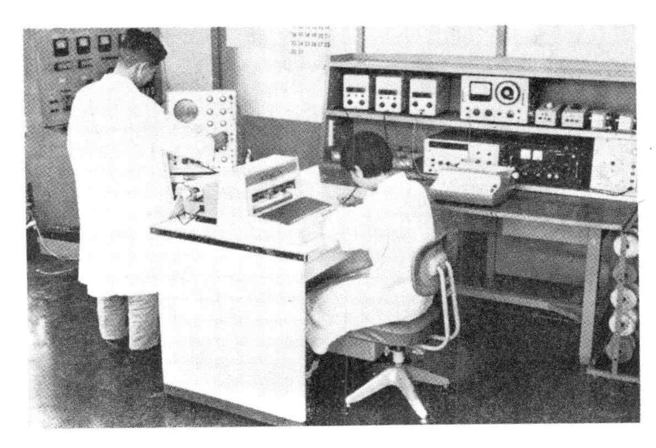

3.8a and b Durability test for Brother typewriters

3.9 Brother Typewriter Model 1350 of 1972 has a zip case, two-colour ribbon, and weighs 11 lb. It also has a pre-set 9 stop tabulator, automatic power spacer, is available in one type style only, and is approximately 11 pitch

With the story of the Brother typewriter, history is repeated! It will be remembered that the first Remington typewriter resembled the sewing-machines for which the company was already famous. Similarly the Brother Company had begun life in the world of sewing-machines.

In 1908 Kanekichi Yasui left his job at the Japanese Army's arsenal to start his own business—the Yasui Sewing Machine Company—in Nagoya, repairing sewing-machines and making spare parts.

He trained his eldest son, Masayoshi, to take over the business, and eventually in 1925, other sons entered the business, and the name of the company was changed to YASUI.

By this time, Masayoshi had not only gained valuable experience of repairing sewing-machines but had studied at night school to acquire the basic knowledge needed to become a machine manufacturer.

The type of machine that Masayoshi had chiefly been dealing with was the chain stitch machine used for the manufacture of straw hats, and it was this kind of machine which he launched on the market in 1928 under the brand name of 'BROTHER'. All six brothers entered the business and the husbands of their four sisters also joined.

A shrewd business man, Masayoshi soon realized the importance of diversifying, and he entered the field of manufacturing woollen yarn knitting machines. In 1934 the firm's name was changed to Brother Industries Limited.

During World War II the firm was asked by Mitsubishi Heavy Industries to manufacture aeroplane engines. Having gained experience of this type of work, and having noted, on a visit to the United States in 1950, the marked increase in the use of electrically operated equipment in American homes, Masayoshi set about producing electrical goods for the Japanese market.

He realized that the age of the treadle machine had gone, and one of the first electrical items to be manufactured was an electric sewing-machine. Further diversification took place and in 1954 the firm took a close look at the possibility of manufacturing typewriters.

The typewriter had rapidly become an important item of office equipment in post-war

Japan and at that time, like the sewing-machine at the beginning of the century, was completely monopolized by foreign capital.

The typewriter is a more complex precision machine than a sewing-machine, and it was some time before Masayoshi felt confident of being able to manufacture a good, original model. Production of typewriters began in 1961, and in 1966 work began on their Mizuho Typewriter plant which was completed a year later.

Today 92 per cent of all Brother typewriters are earmarked for export.

Figures for 1965 show that of the many different articles produced by Brother Industries, the sewing-machine still leads the way and represents 43·67 per cent of their total output; this is followed by 20·34 per cent for electric appliances, 16·99 per cent for knitting machines, whilst typewriter production represents a mere 12·77 per cent of their grand total.

'ERIKA'—'IDEAL': DRESDEN, EAST GERMANY

The firm of Seidel and Naumann of Dresden, Saxony, East Germany, set up in business about 1870. They had a machine workshop and a good technical knowledge and background. Naumann's stepfather was a saddler and his uncle made clock cases for the Court of Saxony. Looking round for a profitable product to manufacture, they decided on sewing-machines.

In 1872, they obtained a licence to manufacture 'Singer' sewing-machines and sought first to capture the German market and then the export market. This was mainly Naumann's idea but apparently Seidel did not agree with this and considered the risk was too great, so he left the business.

In 1887, the firm began diversifying and started making bicycles, and in 1892 they manufactured instruments for measuring the speed of locomotives. In the same year the first steps

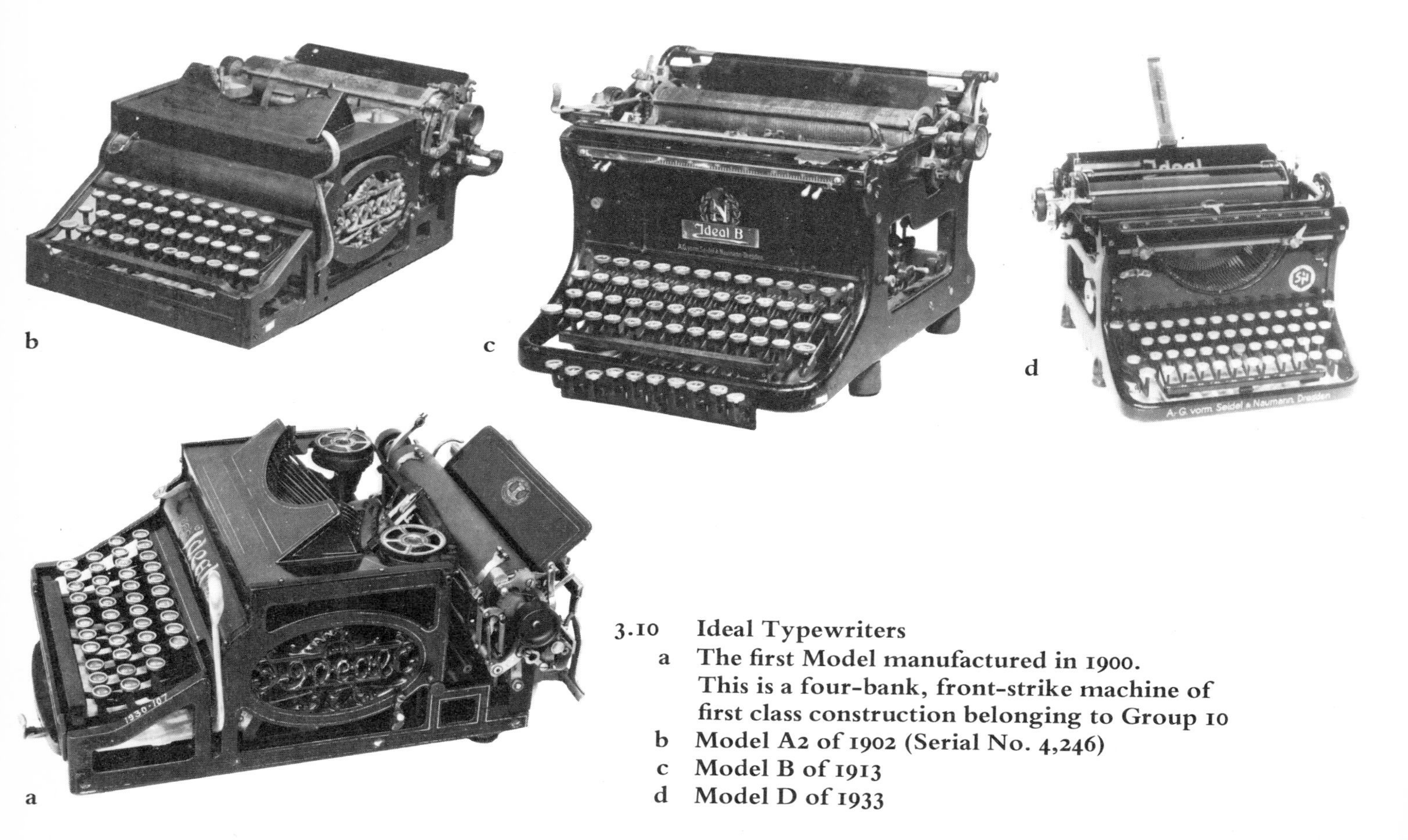

3.10 Ideal Typewriters
- a **The first Model manufactured in 1900. This is a four-bank, front-strike machine of first class construction belonging to Group 10**
- b **Model A2 of 1902 (Serial No. 4,246)**
- c **Model B of 1913**
- d **Model D of 1933**

3.11 Erika Typewriters

a A portable model of 1910, This is a three-bank, double shift machine belonging to Group 10. It was also sold named as Standard Folding, Gloria, Bijou and Bijou Folding

b Model 5 of 1927. A four-bank machine of Group 10 which was also sold named as Ideal, Gloria and Bijou

c Model M of 1936. A four-bank machine of Group 10

d Model 8 of 1948. A four-bank machine of Group 10

e Model 30/40 of 1965. A four-bank machine of unit construction with or without tabulator belonging to Group 10

f Model 32/42 of 1969. A four-bank machine of unit construction belonging to Group 10, and an improved version of Models 30 and 40

g Model 33/43 of 1972. This is the latest and most improved version of the machine, Model 33 has no tabulator and Model 43 has a keyset tabulator

h Model 41 of 1972. This machine is basically the same as Model 43 with the addition of a 13 in. carriage

3.12 View of the exterior of the Facit-Halda Aktiebolag factory at Svangsta, Sweden

were taken to mass produce typewriters. Their first typewriter, the 'Ideal' appeared in 1900 and was a semi-front-strike design. They were eminently successful with it for it sold extremely well in world markets.

In 1913 an entirely new 'Ideal' typewriter was produced, this being a front-stroke machine, but World War I was close at hand and when hostilities began, the firm was soon making shell fuses, rifle barrels and firing pins.

After the end of World War I, they continued with their production of typewriters, sewing-machines, and bicycles. The first three-bank 'Erika' portable was also sold in various countries as the 'Bijou'. It is said that the 'Erika' was named after Erika, Naumann's daughter. Production of the new standard machine continued and later a very high quality four-bank portable was produced.

(cont. page 104)

See Facit page 104

a

b

3.13 Halda Experimental Model Typewriters of 1890/1900

a Model with vertical typebars, blind typing line, three position carriage shift, ribbon and downward typebar travel

b Model with vertical typebars, three position carriage shift, and ink pad. It also has forward, upward, and downward typebar travel

c Model with horizontal typebars, visible typing line, and reverting to ink pad. It also has upward, forward, and downward typebar travel (grasshopper)

3.14a

3.14b

3.14c

Halda Typewriters (*see* Facit page 104)

a Model of 1900–5 with vertical typebars, blind typing line, double keyboard, ribbon and keyboard, ribbon, and paper bail with scale. There were approximately ten of these machines manufactured

b Model 4 of 1905–14 with typebars arranged in a circle, typebar travel upwards against overhead platen, blind typing line, and ribbon. There were approximately 100 machines manufactured

c Model 8 of 1914–18. This was their first machine of modern design and it had a two position carriage shift, tabulator, visible typing line, automatic ribbon lift, and segment with interchangeable typebars. The escapement was actuated by a key lever and it was the first machine with an escapement wheel and two dogs

d Model 10 of 1918–20. This was their first machine with the escapement actuated by typebars. It had a tabulator rack with column locator, shift with shift lock, exchangeable platen, outsize moulded keys with ring and glass, and also exchangeable typebars

e Model 12 of 1924–7. This was their first machine with a noiseless carriage return and the chief difference between this and older models was in the appearance. There were 1,500 machines manufactured while the old company was in liquidation

f Halda-Norden model of 1929–41. This was the first machine made by the present company and it was manufactured during the initial years under licence to AB Skrivemaskinefabriken Norden of Copenhagen. It was later followed by models 2, 3, 4, and 5 with minor modifications. Marketed in the U.K. as Oliver 20

g Model 6 of 1941–57. This is largely the same as the previous Halda-Norden model but it has an entirely new exterior with green finish and a decimal tabulator

h Model 7 of 1953–8. This is of completely new construction; it has block, keyset and decimal tabulators and also a tube and ball-bearing mounted carriage. There is a new keyboard with forty-five keys

i Model P of 1948–59. This is a completely new portable typewriter, comparable in appearance and performance to large portables on the market. It has green housing

When World War II started, the firm, which by then had produced a million typewriters, had to use their factory almost exclusively for producing armaments of various kinds, presumably similar to those manufactured during World War I. Two large scale air attacks were made on the factory, the second of which took place right at the end of the war in April 1945, destroying about 75 per cent of the production area.

Production of typewriters was resumed immediately after the war and the firm concentrated on the manufacture of portables. By 1948 the 'Erika 8' was in production, to be followed by the 'Erika 9'. Further improvements in the model came in 1952 with the 'Erika 10', and in 1954 with the 'Erika 11'. In 1963 Models '10' and '11' were replaced by Models '14' and '15'.

In the middle of 1965, a completely new product of unit construction was developed. This was the '30/40'. The whole assembly was rationalized and the factory was reorganized with modern production assembly methods.

The latest models are the '33' and '43'. Like all typewriters manufactured in the Eastern Zone of Europe, they are produced by the State and the workers participate in the profitability of the enterprise. The 'Erika' is marketed under its own name which is encouraged as it is a first class typewriter. In England, however, it is sold as the 'Boot 33' and '43', and in other countries under different names.

Table 3

PRODUCTION OF BIJOU, ERIKA OR IDEAL TYPEWRITERS

Bijou, Erika or Ideal Portable		**Ideal**	
No. 1		**Model A**	
To 6,740 ..	1910	To 103,386 ..	1900—1916
No. 2		**Model B**	
6,741 to 63,900 ...	1911—1923	Under 30,000 ...	1912—1926
No. 3		**Model C**	
63,901 to 78,900 ...	1923—1925	30,001 to 100,000 ...	1920—1926
No. 4		**Model D**	
78,901 to 89,999 ...	1926—1928	100,001 ...	1926—1929
		140,001 ...	1930—1932
Nos. 5, 6, M. and S. Models		260,001 ...	1933—1934
To 100,000 ...	1927—1928	285,001 ...	1935
100,001 ...	1929—1932	300,001 and over ...	1936
150,001 ...	1933—1934		
200,001 ...	1935		
300,001 ...	1936		
600,001 ...	1937		
700,001 ...	1938		
800,001 and over ...	1939		

At 22,000 all **Ideal Models** began to be numbered consecutively. Some Model B's are also known as Model No. 5.

THE 'FACIT' ORGANIZATION, PREVIOUSLY 'HALDA': SWEDEN

Facit-Halda Aktiebolag was founded in 1887 by Mr. Henning Hammarlund. The Company was originally called 'Halda Fickursfabrik' and dealt with the manufacture of watches; this was mainly due to the fact that Mr. Hammarlund was a watchmaker by profession.

Mr. Hammarlund spent a great deal of time training and learning techniques in various countries, including the United States of America. He found there was not really very much scope in Sweden for watchmaking and soon became interested in making typewriters.

In 1890 he wrote a small book entitled *The Typewriter* in which he gave general guidance about the machines he was manufacturing together with various models and patented inventions of the Company. It also gave instruction in typing, which was very similar to the basic method of 'touch typing' as it is taught today.

He expanded the typewriter side of the business and eventually concentrated all his efforts and capital on the manufacture of these machines, so that in 1938 Halda were able to join with Facit AB—a Swedish company with sales organizations all over the world and an international reputation to uphold. With this group behind them they had plenty of scope for improving their machine and devising new methods.

a

b

c

d

e

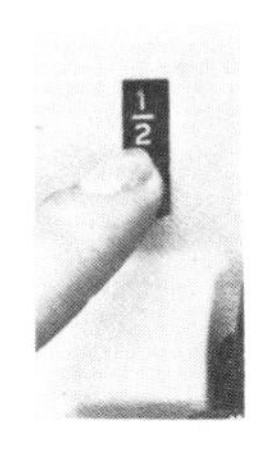

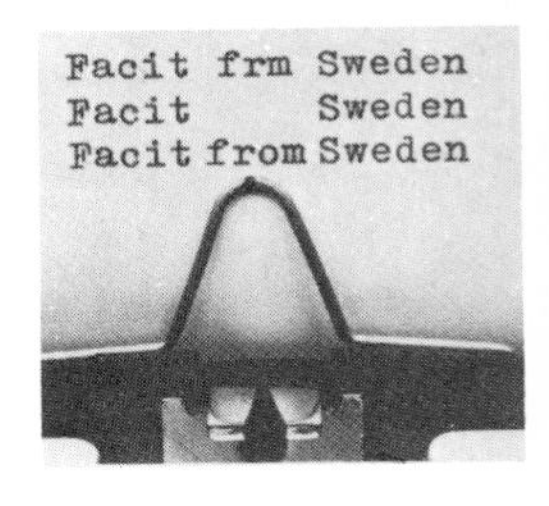

Half-step correction

3.15 Facit typewriters

a Model T1 of 1957–60. This machine has light alloy housing cast in one piece and finished in Viking Grey and the exterior was designed by Sigvard Bernadotte. It has a tube and ball bearing-mounted carriage which is easily lifted out; there are pre-set, keyset, and decimal tabulators and also a new ribbon system with DIN spools

b Model P1 of 1958–64. This is an improved version of the Halda P machine. It is in Viking Grey with the exterior designed by Sigvard Bernadotte. There are keyset and block tabulators with seven pre-set stops, and also a tube and ball bearing-mounted carriage

c Model T2 of 1960 (from Serial No. 231,601). This has lightly modified housing with a smooth finish. There are ribbed typebars, larger typebar journalling, graphite grey keys, a paper support and an erasure plate

d Model P2 of 1964 is an improved design of Model P1. In 1967 there was another design and the colour was changed to beige

e Model 1730. This is the latest 1972 Model and has a half-step correction. In order to correct an error such as a 'dropped' or an extra letter, one erases the word and uses the typing position shift key (which should be depressed when making the correction), and this moves the typing position a half-step sideways

One of the chief designers at this time was Sigvaard Bernadotte. In association with Bjorn he spent a great deal of time on research into the problems of materials, heat treatment, surface treatment, and modern design.

Roughly 30 per cent of the Company's output is sold in Sweden and the remainder is exported throughout the world by the sales organization of Facit.

With the increase in production and entrance into the standard, electric, and portable markets, additions and modifications were built onto the original factory at Svangsta and two new factories were established at Brakne-Hoby in 1959 and Solvesborg in 1961.

The Brakne-Hoby factory specializes in portable typewriters, leaving the Solvesborg factory to concentrate on the improvement of light metal goods including the casing for machines.

a

b

d

e

3.16 Facit Electric Typewriters

- a **Model ET1 of 1959–61. This is a completely new electric typewriter of a smooth grey colour. The typing mechanism features a positive drive shaft of steel and there are keyset, pre-set, and decimal tabulators. The motor switches off automatically and there is a low keyboard with four repeating keys**
- b **Model ET2 of 1961–3 is an improved version of Model TE1 and is approved by the Swedish Bureau for Testing Electrical Equipment. There are six repeat actions and no automatic switch-off. There is a paper support and an erasure plate**
- c **Model ET2 of 1963–9. This shows an alternative view**
- d **Model ET3 of 1963–9. This is an improved model of ET2**
- e **Model 1820 of 1969. This is a completely new electrical typewriter equipped with several new features such as pre-programmed tabulator, separate line spacing at any carriage position and separate carriage return. It also has a textile and carbon ribbon which is adjustable by a lever on the keyboard and is built for systematic typewriting**

The factories now employ approximately 1,600 people. Of these 1,400 actually work in the factories and the rest are outworkers or women at home who do the equivalent of fifty to sixty full-time workers' jobs. There is a great deal of light work which can be accomplished in this way. Facit-Halda takes advantage of this to the full and its Planning Department has transport which takes the work to these people and then collects it later.

Table 4

PRODUCTION OF FACIT AND HALDA (NOW FACIT) TYPEWRITERS

Facit Standard Manual

T.I. Standard	
Up to 117,656	1958
117,657	1959
158,951 to 170,499	1960

T.2 Standard	
170,500	1960
192,330	1961
238,858	1962
272,096	1963
332,222	1964
379,535	1965
424,696	1966
450,851	1967
511,748	1968
607,037	1970
661,228	1971

Facit Portable

T.P. 1 & T.P. 2	
Up to 205,900	1958
205,901	1959
223,054	1960
260,995	1961
300,760	1962
305,989	1963
329,364	1964
366,335	965
531,551	1966
537,321	1967
628,049	1968

Facit Electric

Model ET3	
Up to 3,110	1960
3,111	1961
12,360	1962
20,300	1963
32,980	1964
44,953	1965
71,032	1966
81,526 to 109,230	1967

1620 Series	
Up to 743,285	1969
743,286	1970
806,059	1971

1820 Series	
Up to 205,946	1969
205,947	1970
237,375	1971

Halda Standard Manual

Model 6.K	
202,000	1950
222,001	1951
255,001	1952
283,001	1953
290,001	1954
294,001	1955
296,001	1956
300,001 up	1957

Model 7 Star	
Up to 7,600	1953
12,000	1954
15,001	1955
26,001	1956
39,001 up	1957

Halda Portable

Serial	Year
10,400	1951
12,201	1952
29,501	1953—1954
48,001	1955
72,001	1956
91,001	1957
116,001 to 130,000	1958

The Company also provides homes and social amenities for its workers. They have a Community Centre established at Svangsta. A training school is available as the Company is most progressive and desires a general increase in the level of knowledge of its workers. They train persons to fill qualified positions and perform specialized tasks.

There is a tradition in the Company that anyone who works hard and gains qualifications in the training school can reach the very top of his profession, even though he may have started on the very lowest rung of the ladder.

The first typewriter the Company produced was an experimental one in 1896. A great deal of research and development followed this until their first typewriter was ready for the market in 1914. However, war interrupted their programme and for a while there was very little activity. In 1927 fresh capital was invested and thus the Company was enabled to continue its programme, producing a new machine in 1929 called the Halda-Norden. Most Facit-Halda typewriters use Ransmeyer type.

They did not enter the field of portable typewriters until as late as 1947, but from these beginnings has grown the organization as it is today. As the Facit Concern, they form one of the leading Swedish companies with sales organizations in 140 centres in Sweden and 130 countries throughout the world. The factory at Svangsta is situated in the most beautiful surroundings. Photographs of this factory and the various models made are illustrated (*see* Figures 3.12–3.16).

Facit is licensed for production in Poland under the name of Lucznik.

THE GODREJ 'ALL INDIA' TYPEWRITER

Growth and progress from a small beginning is the story of Godrej. Established in 1897, they have an enormous production of steel goods including refrigerators, vault doors for strong rooms, steel cupboards, steel chairs, and fork-lift trucks, aluminium and steel windows, doors, ventilators, and partitions, together with large quantities of soap and chemicals for home consumption. They employ over 6,000 workers in their machine tool plant alone.

The company is most progressive in outlook, providing schools for technical education and free meals and uniforms for all students and employees, as well as comfortable workers' houses in garden settings. The factory is situated on the main industrial highway of Greater Bombay, between Santa Cruz International Airport and the Atomic Research Centre.

In 1955, they commenced manufacturing typewriters and produced the original Godrej 'All India' which was the 'Woodstock' 1955 model, i.e. R. C. Allen 600/700 series. Between 1958 and 1967 the Godrej Model '12' was produced. This was achieved in co-operation with the East German 'Optima' factory. *See* the photographs opposite.

After 1967, the firm severed their connection with the East German firm and produced the current machine known as the Godrej Model 'AB'. It is of modern design with fast feed, produced in eight type faces and three carriage lengths. The factory mainly supplies the Indian and surrounding local markets.

At the time of writing it is difficult and expensive to import European or American typewriters into India due to a shortage of foreign exchange.

THE HAMMOND TYPEWRITER AND THE VARITYPER OFFICE COMPOSING MACHINE: U.S.A.

A young Civil War correspondent read his own dispatches in the newspapers and was dismayed and angered at the garbled way in which they had reached print. His handwritten reports had been completely misread and the printed stories were a source of embarrassment to him. There was little he could do about it; the urgency of getting his stories to the telegraph

a

b

c

3.17 Godrej Typewriters
- a This machine was produced in India in 1955 and was modelled on the Woodstock
- b Model 12 of approximately 1958–67 was modelled on the Optima
- c Model AB of 1972 is the current machine

3.18 Godrej India Typewriter assembly

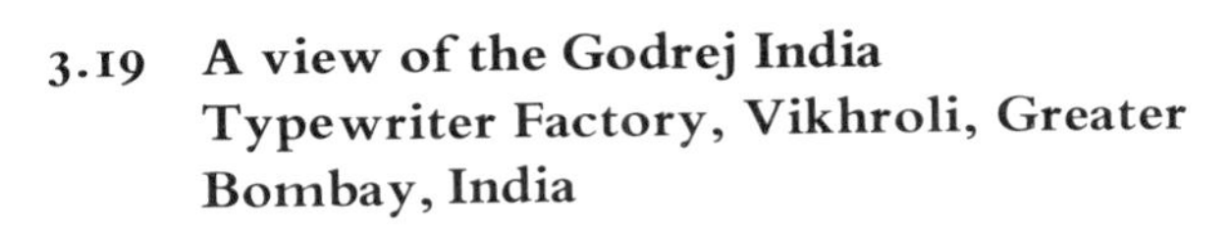

3.19 A view of the Godrej India Typewriter Factory, Vikhroli, Greater Bombay, India

3.20 Hammond Typewriters made in U.S.A. (all of Group 1)

a Model No. 2 of 1893. This is of the swinging sector design with a two-row circular keyboard

b Model No. 2 B of 1893 has a square keyboard

c Model No. 12 of 1905

d Multiplex A Model of 1910

e This Folding Model of 1921 is portable, and has a folding square keyboard

f Multiplex B Model of 1923 has a three-row square keyboard

stations afforded little time for exercises in penmanship. Resignedly, he consigned his hastily written reports to the indifference of the telegraphers and typesetters. But dissatisfaction often breeds creativity, and during lulls in military action he daydreamed, conjuring up images of 'miracle machines' that would write mechanically with printer's type. The writing would be so clear that even a careless telegrapher could transmit his stories without error. The Civil War dragged on. James B. Hammond, the young correspondent, continued to see his stories misinterpreted and misprinted.

James Hammond's dreams of a miraculous writing machine crystallized to the extent that he began to put them down on paper in the form of sketches. That was the beginning of the Hammond Typewriter, one of the earliest practical 'Writing Machines'.

The dream bore fruit . . . James Hammond produced his first commercial model in 1881, and for its day it was a very advanced machine. However, the idea of typing by touch rather than sight had not occurred to Hammond or any of his contemporary inventors. Since speed was not the primary consideration, it was thought sufficient that the operator could easily locate the characters on the piano-like keys and strike them to print the desired letters. The keyboard was semi-circular in configuration . . . the more popular straight-line keyboard was to be adopted in later models. Two important ideas saved the Hammond from the early demise of most of the 'Writing Machines' of that era—type that was quickly and easily changeable, and a system of uniform impression that did not depend on the force with which the keys were struck.

While never an outstanding commercial success, the unique qualities of the Hammond machine endeared it to a small but devoted following.

After Hammond's death in 1913, his company drifted into obscurity, to be revived in the early 1930s by Ralph C. Coxhead who saw the machine becoming an adjunct to office duplicating. Under his inspired leadership, development followed development in quick succession, and the product that emerged became the world's first cold type composing machine. Gone forever was the name 'typewriter', and in its place the new nameplate bore the name 'VariTyper, the Office Composing Machine'.

In its new role, the VariTyper machine met immediate success in the graphic arts and in inplant composition and duplicating departments, providing business and industry with new and less expensive means of communicating in print.

VariTyper joined the total mobilization of World War II, serving where conventional typesetting machines were of little use because of their weight, bulk, and complexity. It served aboard ship, and with the armies abroad; and it served industry as it supported the war effort on the home front. Significantly, too, VariTyper helped write '*finis*' to the war as it composed the Instrument of Surrender which was signed by the Japanese on board the battleship *Missouri*.

It is true that swords are often beaten into ploughshares. As the Hammond Typewriter was born of a wartime need and found its way into peaceful pursuits, the VariTyper Machine,

a

3.21 VariTypers
a Hammond VariTyper of the mid 1930s. This model is an early Coxhead machine (Serial No. B 37,103)
b This is an early Model of 1927–8, it has a three-bank straight keyboard, was made in the U.S.A. and is of Group 1
c Model No. 1010 Light Touch of 1968

b

c

too, found its way from the conflict of World War II into the fight for peace that followed. VariTyper Machines, today, are found in practically every country of the world, often where the printed word would otherwise be too costly or too difficult or too slow to be put into circulation.

VariTyper's mainstay, however, is and always has been its place in thousands of plants and offices the world over, playing its role as a part of the graphic communications team, helping people communicate.

The VariTyper is a machine for composing typed matter for 'offset' printing. This method of type composition was pioneered and developed by VariTyper over the years and has come to be known as 'Strike On' Typography. It owes its name to the fact that professional quality typography is produced by a simple typing operation, just as with an ordinary typewriter.

The ordinary typewriter rules itself out for professional uses as it is limited to only one type face and spacing alignment, making it impossible to 'justify' columns (i.e. create a uniform left and right margin). The process of creating an 'offset' duplicating master by hand-setting founders type faces or by linotyping is prohibitively expensive from the standpoint of the hourly wage paid to the printer, and the time required to convert the text into an 'offset' master.

Table 5

THE VARITYPER

1933 . . . We **introduced** the VariTyper Machine—the machine that fathered an industry—'Cold Type'. Its changeable sizes and styles of type revolutionized inplant duplicating.

1934 . . . The **carbon ribbon** was added to give cold type the finest impression possible. We originated it, and since then it's been adopted for use on many typewriters.

1937 . . . Cold type gained new professionalism with the addition of **automatic justification** to square the margins just like in conventional typesetting.

1947 . . . **Differential Spacing** raised cold type to new levels of typographic excellence. At the same time, a new and superior method of automatic justification was introduced. It's still the best in the field.

1950 . . . A versatile machine became even more versatile with the introduction of the **Automatic Forms Ruling Device.** And VariTyper is still the only cold type machine that can rule such a variety of lines and leaders with such precision.

1953 . . . More flexibility was introduced with the **changeable coder**. The typographic (and language) flexibility it provides is still unmatched in the industry. At that time we also introduced the **Line Spacing Device** that permits accurate line spacing from ½-point to 18 points in ½-point graduations.

1956 . . . The Line Spacing Device was further refined to permit accurate feed of the paper down or up.

1963 . . . The **VariTyper 660** brought new simplification to many typographic functions along with newly updated styling.

1967 . . . The **VariTyper 720** further automated many of the operating features resulting in faster, easier operation.

1968 . . . And now the new **VariTyper 1010** changes the whole shape of cold type as it adds the speed and ease of operation of a *powered keyboard* for greatly increased net output. At the same time, many parts have been completely re-engineered to bring new quality to cold type.

The simplicity and low cost of the VariTyper method has made possible inplant composition of jobs that formerly could be set only by the printer. This is accomplished by the use of removable sets of type 'founts'. Each fount contains a full set of alphabetic and numeric characters of a particular style of type, and there is a VariTyper type fount for practically every style of type.

The VariTyper machine also has the ability to vary the spacing between the typed characters according to the requirements of the text being composed. This ability, along with the ability to justify columns results in an end product that is professionally correct. Only an expert can distinguish between an offset master prepared by VariTyper and one prepared from cold type (either hand-set or linotyped).

A number of 'VariTyper' machines are now produced by 'Addressograph-Multigraph Division'. They have gradually been improved throughout the years. Numerous models are available for specialized purposes. As a general rule they are not used in offices as ordinary typewriters, but as composing machines, developed from a very successful typewriter that operated and sold *as a typewriter* from 1880–1920, a matter of forty years. All present 'VariTypers' are electric; the 'Hammond' manually operated.

We are indebted to Addressograph-Multigraph Ltd. for the information provided here and for the quotation reproduced as Table 5.

HERMES: SWITZERLAND

The Hermes typewriter was produced by the firm of PAILLARD—a firm which had begun life at the beginning of the nineteenth century by manufacturing musical boxes in Sainte-Croix, which was then only a tiny village, perched on the heights of the Jura mountains in Switzerland.

The industry started by Moïse Paillard grew and became increasingly important under the management of his descendants. In 1875 the first factory was built at Sainte-Croix. By the end of the century the firm had begun to diversify, and shortly after the invention of the phonograph, Paillard proceeded from musical boxes to sound-reproduction equipment, and as time went on more and more sophisticated equipment left their factories—gramophones, record-players, and radio receivers.

The Paillard management first took an interest in typewriters in 1913, although manufacture did not start until 1920. In that year a new factory was built at Yverdon, and this factory was soon used solely to produce the Hermes typewriters. So successful were they that they had to extend their factory on many occasions to cope with the demand for their machines.

It is interesting to note that Paillard right at the outset had started research into the model most difficult to develop—the standard office typewriter. The first model he produced was devised from a model shown by a French inventor, but important modifications were made, especially on the carriage.

The first Hermes typewriter never passed the prototype stage, but, incidentally, it was the first in the world to be fitted with an automatic tabulator. 'Hermes 2' was launched commercially in 1923, but proved too expensive in its manufacture, and it was soon outdated by the more advanced models of some of their competitors. Consequently only 1,000 machines of this model were made.

The 'Hermes 3' was launched in 1927, and in 1928 the 'C' (15-inch) carriage was introduced, to be followed in 1930 by the 'A' (10-inch) carriage and in 1932 the 'D' (18-inch) carriage. The 'Hermes 4' closely resembling the 'Hermes 3', appeared in 1934.

In 1933 the company launched their first portable—the 'Hermes 2000', and with this machine they began to penetrate numerous foreign markets.

Two years later, in 1935, the 'Hermes Baby' was launched. It was invented by a Mr. Prezioso, and had all the features of a really portable machine, weighing less than 8 lb, and it was no higher than a big box of matches!

Between then and the end of World War II the company introduced new models as follows:

1936 'Hermes Media'
1937 'Hermes 5'
1939 'Hermes 2000' fitted with the first automatic margins
1940 'Hermes 2000 Jubilee'
1940 'Hermes Baby Jubilee'
1943 'Hermes 6'

From selling some 250 machines a year in 1926, Paillard increased their sales to 1,100 in 1928, to 15,500 in 1935 and more than 35,000 in 1936.

In 1938 Switzerland, thanks to Paillard, had become, with 42,000 machines, the third typewriter exporting country, behind only the United States and Germany.

The Company continued to prosper after the war, and in 1948 they launched the 'Hermes Ambassador', a manual office typewriter, in which were found several revolutionary advantages, such as: incorporated notebook holder, automatic paper insertion and ejection, electric return of carriage, and interline spacing.

Other post-war dates of importance are:

1953: Launching of the 'Hermes 8', a manual office typewriter which incorporated such innovations as: vertical carriage rails, setting and clearance of tab stops by a single lever, incorporated notebook holder.

1954: Launching of a new model 'Hermes Baby', equipped with a larger diameter platen (31·5 mm).

1958: Launching of the 'Hermes 3000', introducing numerous innovations including: lightening margins visible in front of the sheet of paper (first typewriter in the world with this system), a control board grouping all the service keys, an integrated base plate and cover to facilitate handling and carrying.

1959: Launching of the 'Hermes Ambassador Electric'. The system driving the typebars differs from the classical systems (fluted power shaft and roller power shaft). The dual ribbon version (fabric and carbon) of this model constitutes an innovation; a single selector knob enables switching from the carbon to the fabric ribbon (single or twin-coloured), and vice-versa, depending on what type of work has to be done.

1962: Construction of the factory at Säckingen, Germany, to manufacture 'Hermes Baby' typewriters.

1964: Launching of the 'Hermes 9' office manual typewriter. The typing system was specially developed to give a light and rapid touch.

1965: Construction of the Santo Amaro factory to manufacture 'Hermes Baby' typewriters.

1967: The three millionth 'Hermes Baby' came off the production line.

1968: Launching of the 'Hermes 10' office electric typewriter; of convenient size, yet incorporating all the advantages of a bigger machine.

In the 1930s the firm had also been manufacturing movie cameras—the Bolex cameras, and after World War II, the success enjoyed by the Hermes typewriter and the Bolex movie equipment persuaded the management to abandon production of gramophones and radio sets and concentrate on the two principal product lines. In 1970 Paillard signed an agreement with the Austrian group, EUMIG, under which Eumig gradually took over the production of the Bolex equipment.

Thus today typewriters—and as in many other firms their 'logical' developments—calculators and systems machines, constitute the essential activity of Paillard. The machines are made in nine factories, situated in Switzerland, France, and Germany; and sold under the brand names, HERMES, JAPY, and PRECISA.

b

c

e

f

i

h

3.22 **Hermes Manual Standard Typewriters**

- a Model No. 1 of 1920
- b Model No. 2 (Serial No. 1,300 in 1923 to 2,000 in 1926)
- c Model No. 3 (Serial No. 2,000 in 1927 to 12,363 in 1934)
- d Model No. 4 (Serial No. 13,001 in 1934 to 18,177 in 1936)
- e Model No. 5 (Serial No. 500,001 in 1937 to 539,600 in 1943)
- f Model No. 6 (Serial No. 539,601 in 1943 to 620,200 in 1953)
- g Ambassador (Serial No. 736,201 in 1953 upwards)
- h Model 8 (Serial No. 8,000,001 in 1954 to 8,048,701 in 1959 upwards to 8,088,000 in 1962)
- i Model 9 (Serial No. 8,151,200 in 1966)
- j Ambassador Version 3 (Serial No. 1,000,000 in 1968 upwards)

3.23 Hermes Electric Typewriters
a Compact Model 10 (Serial No. 2,000,000 in 1968
,, No. 2,001,300 in 1969
,, No. 2,012,200 in 1970
,, No. 2,037,000 in 1971)
b Ambassador (Serial No. 845,324 in 1959)
c Ambassador (Serial No. 897,692 in 1962)
d Ambassador (Serial No. 4,119,648 in 1971)

Hermes Portable Typewriters (*see* opposite)

3.24i

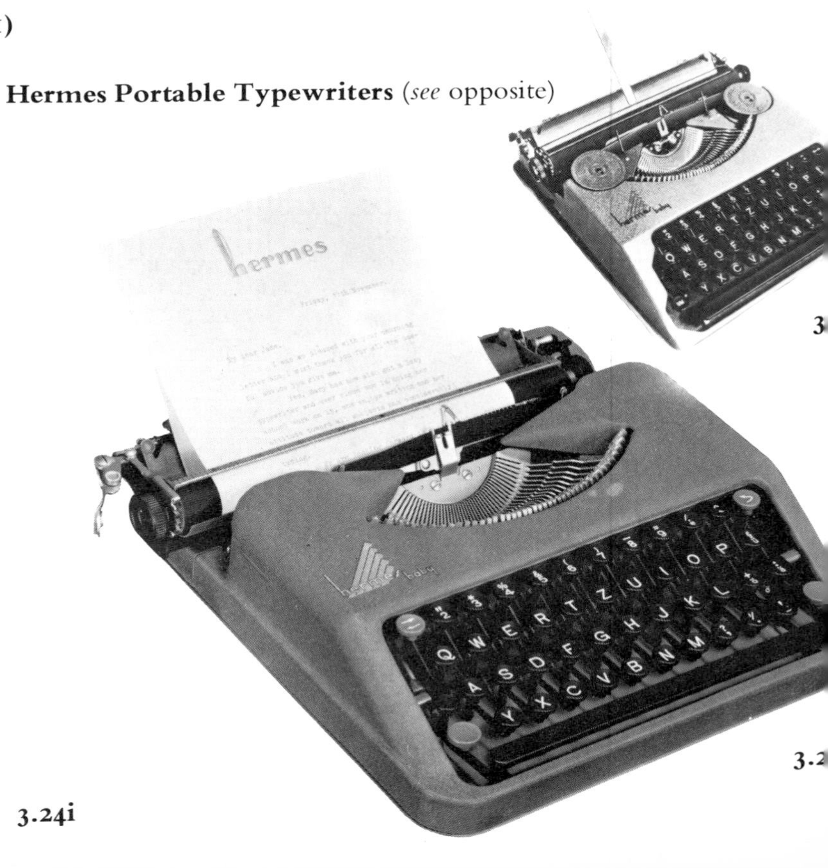

3.2

b

c

d

g

3.24 Hermes Portable Typewriters

a Media Model which was sold in the U.K. as the Empire Junior. (Serial No. 30,001 in 1936 to 33,701 in 1937)

b Model 2000 (Serial No. 20,001 in 1933 to 33,700 in 1936)

c Model 2000 was also sold as the Empire junior. (Serial No. 360,001 in 1940 upwards)

d Model 3000 (Serial No. 3,112,500 in 1962
,, ,, 3,165,500 in 1963
,, ,, 3,222,000 in 1964
,, ,, 3,279,000 in 1965
,, ,, 3,336,000 in 1966
,, ,, 3,391,500 in 1967
,, ,, 3,455,000 in 1968
,, ,, 3,518,400 in 1969
,, ,, 3,571,900 in 1970)

e Model 3000 (Serial No. 7,070,052 in 1971)

f Hermes Baby which was sold in the U.K. as the Baby Empire and was manufactured under licence in West Bromwich, England. (Serial No. 60,001 in 1935 to 173,400 in 1939)

g Hermes Baby (Serial No. 173,401 in 1939 to 5,346,200 in 1953)

h, i Other views of the Hermes Baby
(Serial No. 5,911,000 in 1962
,, ,, 5,961,000 in 1963
,, ,, 6,025,600 and 9,000,000 in 1964
,, ,, 6,079,800 and 9,003,500 in 1965
,, ,, 6,121,000 and 9,018,500 in 1966
,, ,, 6,150,500 and 9,045,500 in 1967
,, ,, 9,089,00 in 1968
,, ,, 9,132,000 in 1969
,, ,, 9,180,600 in 1970
,, ,, 9,228,000 in 1971)

Table 6

HERMES TYPEWRITER MODELS PRODUCED BY PAILLARD, S. A., YVERDON, SWITZERLAND

Hermes Standard

Serial No.	Year
765,201	1955
782,501	1956
800,801	1957
821,801	1958
839,201	1959
855,679	1960
872,934	1961
897,692	1962
921,657	1963
942,352	1964
958,950	1965
985,012	1966
1,010,600	1967
1,027,900	1968
1,048,864	1969
1,067,717	1970
1,091,791	1971

Standard '8'

Serial No.	Year
8,000,001	1954
8,000,601	1955
8,014,301	1956
8,024,201	1957
8,036,501	1958
8,048,701	1959
8,053,719	1960
8,069,384	1961
8,082,636 to 8,100,279	1962

Standard '9'

Serial No.	Year
8,100,280	1963
8,114,495	1964
8,129,318	1965
8,145,967	1966
8,164,899	1967
8,257,339	1968
8,272,314	1969
8,299,256	1970
8,336,824	1971

Hermes Electric
Ambassador Electric

Serial No.	Year
845,324	1959
853,034 to 867,012	1960
4,000,171	1960
4,002,446	1961
4,006,799	1962
4,013,046	1963
4,019,196	1964
4,024,059	1965
4,031,634	1966
4,042,155	1967
4,052,721	1968
4,072,720	1969
4,092,091	1970
4,119,648	1971

Hermes 10

Serial No.	Year
Up to 2,008,830	1969
2,008,830	1970
2,016,364	1971

Model 2000

Serial No.	Year
421,600 to 436,200	1945
436,300	1946
453,701	1947
473,101	1948
494,501 to 510,000	1949
2,000,000	1949
2,000,701	1950
2,016,501	1951
2,037,801	1952
2,062,901	1953
2,085,501	1954
2,107,101	1955
2,130,601	1956
2,155,801	1957
2,187,701	1958
2,219,701	1959
2,233,036 to 2,239,250	1960

Model 3000 (Media)

Serial No.	Year
3,000,001	1958
3,004,101	1959
3,004,349	1960
3,056,176 to 3,099,169	1961
3,106,641	1962
3,159,082	1963
3,222,061	1964
3,282,181	1965
3,324,583	1966
3,394,101	1967
3,459,670	1968
3,517,643	1969
3,568,842	1970
7,061,071	1971

Hermes Portable
Baby

Serial No.	Year
5,773,901 to 5,796,192	1960
5,835,601 to 5,845,724	1961
5,930,381	1962
5,957,011	1963
6,025,231	1964
6,067,500	1965
6,121,400	1966
6,131,350	1967
6,145,861	1968
9,132,173	1969
9,171,326	1970
9,222,042	1971

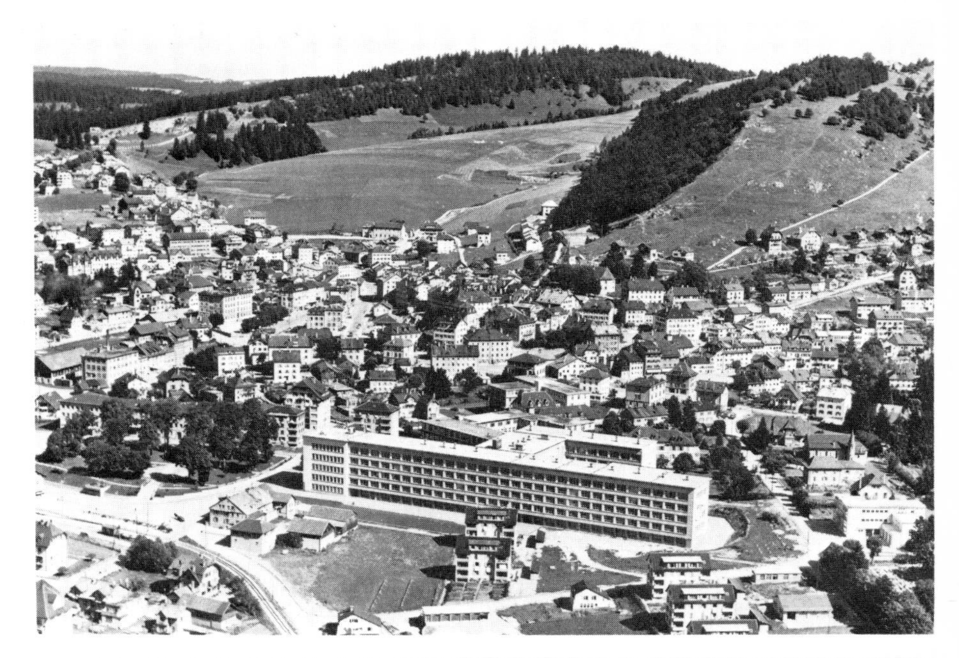

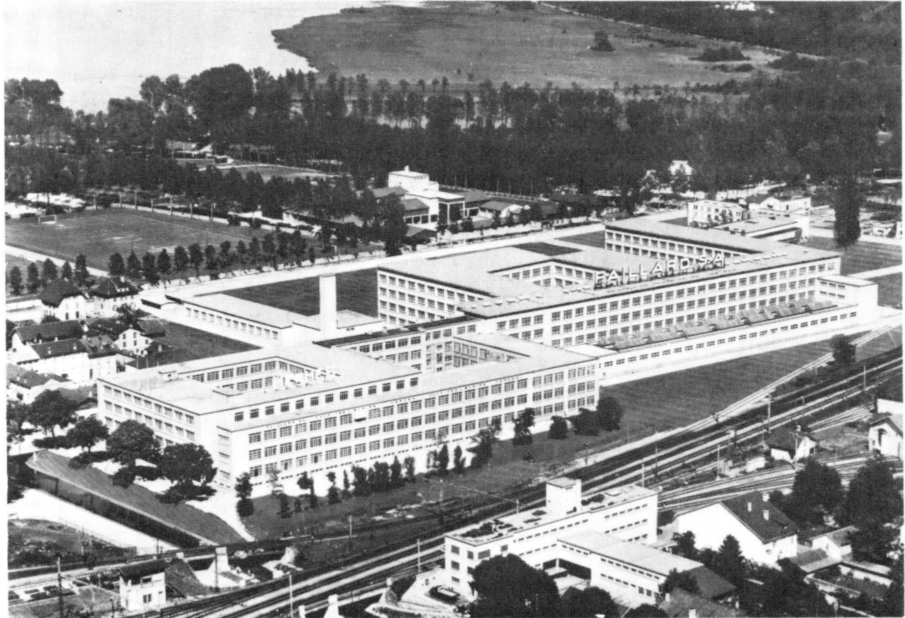

3.25 Views of Hermes's Factories at Yverdon, Switzerland

HERMES Specimens of writing

HERMES Typewriters are available, to your choice with various type styles:

(x = available with this type style; – = not available with this type style)

Code	Type style	Spacing	HERMES BABY	HERMES MEDIA	HERMES 3000	HERMES 9	HERMES 10	HERMES AMBASSADOR	HERMES ELECTRIC	HERMES ELECTRIC VARIA
EC	Economic	Spacing 1,5 mm	–	–	–	x	–	x	x	–
* SC	Script	» 2 mm	x	x	x	–	–	–	–	–
EL	Elite	» 2 mm	x	x	x	–	x	–	–	–
EL	Elite	» 2,12 mm	–	–	–	x	–	x	x	–
HS	Hermes Special	» 2,5 mm	–	x	x	x	x	x	x	–
PI	Pica	» 2,5 mm	x	x	x	x	x	x	x	–
DE	Director-Elite	» 2 mm	–	x	x	–	x	–	–	–
DE	Director-Elite	» 2,12 mm	–	–	–	x	–	x	x	–
DS	Director Special	» 2,5 mm	–	x	x	x	x	x	x	–
DP	Director-Pica	» 2,5 mm	–	x	x	–	x	–	–	–
MI	Modern Imperial	» 2,5 mm	–	–	–	x	–	x	x	–
EP	Epoca	» 2,5 mm	–	x	x	x	x	x	x	–
TE	Techno (Square)-Elite	» 2 mm	–	x	x	–	x	–	–	–
TE	Techno (Square)-Elite	» 2,12 mm	–	–	–	x	–	x	x	–
TS	Techno (Square) Special	» 2,5 mm	–	x	x	x	x	x	x	–
TP	Techno (Square)-Pica	» 2,5 mm	x	x	x	x	x	x	x	–
PP	Small-Pica	» 2,5 mm	–	x	x	x	x	x	x	–
RA	Ocra	» 2,5 mm (only with dual ribbon)	–	–	–	–	–	–	x	–
RB	Ocrb	» 2,5 mm (only with dual ribbon)	–	–	–	–	–	–	x	–
AD	Admiral	Proportional spacing	–	–	–	–	–	–	–	x
AR	Aristos	» »	–	–	–	–	–	–	–	x
AL	Altessa	» »	–	–	–	–	–	–	–	x
AP	Apollon	» »	–	–	–	–	–	–	–	x

*HERMES MEDIA, 3000, available on keyboard Nos. 3002-C, 3002-E, 3002-G, 3003, 3004-B, 3006, 3008-F, 3010-E, 3024-B, 3040-B, 3043-B, 3043-E, 3044-E, 3048-D, 3049-C, 3060-B, 3073-C, 3073-E, 3084-B, 3092-D, 3092-F, 3097-B only.

Definition of signs:

- (white square) Available with this type style
- (grey square) Not available with this type style

Type style	Proportional spacing
Admiral	Inspired by calligraphers and Typography instructors, those who designed these characters knew, admirably well, how to play with the harmonious alternation between black and white: this is HERMES proportional spacing.
Aristos	Inspired by calligraphers and Typography instructors, those who designed these characters knew, admirably well, how to play with the harmonious alternation between black and white: this is HERMES proportional spacing.
Altessa	Inspired by calligraphers and Typography instructors, those who designed these characters knew, admirably well, how to play with the harmonious alternation between black and white: this is HERMES proportional spacing.
Apollon	Inspired by calligraphers and Typography instructors, those who designed these characters knew, admirably well, how to play with the harmonious alternation between black and white: this is HERMES proportional spacing.

	Spacing in mm	
Economic	1,5	The precision of HERMES typewriters and the perfection of their delicately engraved charac-ters, guarantees you a perfect impression year after year. 1 2 3 4 5 6 7 8 9 0
Script	2	The precision of HERMES typewriters and the perfection of their deli-cately engraved characters, guarantees you 1 2 3 4 5 6 7 8 9 0
Elite	2	The precision of HERMES typewriters and the perfection of their deli-cately engraved characters, guarantees you 1 2 3 4 5 6 7 8 9 0
Elite	2,12	The precision of HERMES typewriters and the perfection of their delicately engraved characters, guarantees 1 2 3 4 5 6 7 8 9 0
Hermes Special	2,5	The precision of HERMES typewriters and the perfection of their delicately engraved 1 2 3 4 5 6 7 8 9 0
Pica	2,5	The precision of HERMES typewriters and the perfection of their delicately engraved 1 2 3 4 5 6 7 8 9 0
Director-Elite	2	The precision of HERMES typewriters and the perfection of their deli-cately engraved characters, guarantees you 1 2 3 4 5 6 7 8 9 0
Director-Elite	2,12	The precision of HERMES typewriters and the perfection of their delicately engraved characters, guarantees 1 2 3 4 5 6 7 8 9 0
Director Special	2,5	The precision of HERMES typewriters and the perfection of their delicately engraved 1 2 3 4 5 6 7 8 9 0
Director-Pica	2,5	The precision of HERMES typewriters and the perfection of their delicately engraved 1 2 3 4 5 6 7 8 9 0
Modern Imperial	2,5	The precision of HERMES typewriters and the perfection of their delicately engraved 1 2 3 4 5 6 7 8 9 0
Epoca	2,5	The precision of HERMES typewriters and the perfection of their delicately engraved 1 2 3 4 5 6 7 8 9 0
Techno (Square)-Elite	2	The precision of HERMES typewriters and the perfection of their deli-cately engraved characters, guarantees you 1 2 3 4 5 6 7 8 9 0
Techno (Square)-Elite	2,12	The precision of HERMES typewriters and the perfection of their delicately engraved characters, guarantees 1 2 3 4 5 6 7 8 9 0
Techno (Square) Special	2,5	The precision of HERMES typewriters and the perfection of their delicately engraved 1 2 3 4 5 6 7 8 9 0
Techno (Square)-Pica	2,5	The precision of HERMES typewriters and the perfection of their delicately engraved 1 2 3 4 5 6 7 8 9 0
Small-Pica	2,5	The precision of HERMES typewriters and the perfection of their delicately engraved 1 2 3 4 5 6 7 8 9 0
Ocra	2,5	THE PRECISION OF HERMES TYPEWRITERS AND THE PERFECTION OF THEIR DELICATELY ENGRAVED 1 2 3 4 5 6 7 8 9 0
Ocrb	2,5	The precision of HERMES typewriters and the perfection of their delicately engraved 1 2 3 4 5 6 7 8 9 0

Some special or less popular keyboards are only available in Pica.
Longer delivery may be necessary for certain type styles.

3.27 Hermes Type Styles

CHAPTER 4

History of Typewriter Manufacturers Part II

IBM—INTERNATIONAL BUSINESS MACHINES: U.S.A.

In 1923 a Mr. Russell G. Thompson began work on improving an electrically driven typewriter which had been designed by Mr. James F. Smathers of Kansas. Under the aegis of the North-East Electric Company, good progress was made and a year later, the production of electric typewriters was handed over to the North-East Appliances Company Inc.

In 1929 the name was changed to Electromatic Typewriters Inc. This organization was then completely separated from the North-East Electric Company, although the Directors and shareholders of the two companies were more or less the same.

In the same year the North-East Electric Company was sold to General Motors, but the Electromatic Typewriters remained an independent Company until it merged in 1933 with the International Business Machines Corporation of Rochester, New York.

At that time widespread usage of electric typewriters was not much more than a dream in the minds of a few rather progressive businessmen. However, Electromatic Typewriters had actually made a little money by manufacturing and selling the new machine in the early years of the great depression, but the Company lacked the necessary capital to develop and market the machine on a large scale. Previous attempts by other companies to manufacture and sell electric typewriters had all ended in disaster, and Electromatic and its thirty employees (including four salesmen) would probably have been no exception but for Mr. Thomas J. Watson who had the foresight to acquire this Company to form IBM Electric Typewriter Division. He had left National Cash Registers in 1913 and a year later became President of the Computing-Tabulating-Recording Company which dealt in Hollerith machines. The Company grew and expanded and in 1917, they entered the Canadian market under the name of International Business Machines Co. Ltd. In 1919 the IBM Company had entered the European market and sales had been very good.

IBM then went about putting its development, production, and marketing know-how into perfecting an electric typewriter which would soon start a complete revolution in the typewriter industry.

In 1934 over a million dollars was invested in the new typewriter which was re-designed without any of the operating deficiencies in the previous Electromatic machine. IBM realized that the market had to be convinced that the Electric was safe, reliable, and efficient. A customer

Table 7

ELECTRIC TYPEWRITER MODELS PRODUCED BY IBM

IBM Electric

Any queries on serial numbers of IBM machines should be addressed to the Company.

Model C Standard

91-12 Canada	
92,013	1960
92,328	1961
94,713	1962
95,000	1963
95,701	1964

35-12 France	
30,001	1959
33,696	1960
41,983	1961
50,560	1962
58,052	1963
67,139 to 75,402	1964
76,000 to 81,366	1965

40-12, 71 OR Germany	
20,500	1959
28,106	1960
42,188	1961
60,358	1962
79,881	1963
99,842	1964
117,994 to 138,685	1965

58-12 Holland	
30,001	1959
30,538 to 32,010	1960
31,401	1961
32,620	1963
32,980	1964
35,582 to 43,877	1965

(If the country code, i.e. 35, is followed by 82 instead of 12 it denotes carbon ribbon attachment.)

31-12 U.K.	
30,001	1959
31,001	1960
34,111	1961
36,809	1962
39,204	1963
41,884 to 42,024	1964

OO-12 U.S.A.	
1,000,000	1959
1,240,000	1960
1,385,001	1961
1,510,001	1962
1,643,001	1963
1,780,001	1964
1,943,001 to 2,600,000	1965

Model C Executive

91-42 Canada	
90,327	1959
92,014	1960
94,713	1960
95,000 to 95,700	1962

71 or 42, 40 Germany	
5,136	1959
6,349	1960
9,540	1961
13,954	1962
20,196	1963
28,770	1964
39,253 to 51,999	1965

31-42 U.K.	
50,001	1959
50,348	1960
52,438	1961
54,554 to 56,054	1962
56,055	1963
57,858 to 57,894	1964

Model 72 Selectric

Prefix 58 Holland	
8,001,001	1961
8,001,602	1962
8,013,867	1963
8,028,333	1964
8,046,833 to 8,071,061	1965

engineering operation was established to assure all Electromatic typewriter customers that their machines would be kept in top running condition at all times.

With this programme to support it, the IBM electric typewriter was introduced in 1935 and became the first commercially successful electric to be marketed in the United States. The product line was rapidly expanded in the late thirties with the addition of the Toll Biller, the Manifest Writer, and the Automatic Formswriter, all of which greatly increased the application of the electric typewriter to office procedures.

4.1 An Electromatic Typewriter which was the forerunner of the IBM electric. Manufactured approximately 1933

In 1941 the IBM organization announced a radical breakthrough in typewriter technology. Ever since its invention the typewriter had employed a single-spacing principle which allowed the same letter space width for all characters regardless of size. Inventors and manufacturers had struggled long and hard to develop a simple low-friction carriage mechanism that would single-space without jumping or sticking: by 1941 such a spacing device had been perfected.

IBM engineers spent years researching, developing, and perfecting a mechanism that would measure each alphabetical character in units. In 1944 they announced the first IBM Executive typewriter with proportional spacing. This allowed from two to five units of space per letter and produced material that simulated the appearance of the printed page. After only eight years IBM had successfully solved a problem which had baffled typewriter inventors and manufacturers for nearly eighty years.

It is said that the first of these Executive machines with proportional spacing was presented to President Roosevelt, and his personal letters were typed on this machine. One of these letters was sent to Mr. Churchill who replied that although he realized their correspondence was very important, there was absolutely no need to have it printed!

At the conclusion of World War II, the Armistice documents were typed on an IBM electric typewriter and, later, a similar machine was used to prepare the original UN Charter in San Francisco.

During the World War II years of 1941–45 IBM offered its entire facilities to the American Government for the war effort. They accepted only a nominal 1 per cent profit on articles and produced naval and aircraft fire control instruments, Browning automatic rifles, 30 mm calibre carbines, director and prediction units for 90 mm anti-aircraft guns, bomb-sights, and aircraft super-charger impellers.

In 1944 they produced their first large scale computer—the automatic sequence controlled calculator—which was presented to Harvard University.

It was not until 1946 that normal activities of the Company were resumed and the Executive typewriter was marketed. The Company again invested heavily in typewriter product research and engineering. This resulted in 1948, in the introduction of the completely new model 'A' Standard typewriter which remained the basis of the product line until 1954.

4.2 IBM Electric Typewriters
- a Model AA Series of approximately 1945–6
- b Model 72 of 1961 onwards incorporating the 'Golf Ball'
- c Composer 82 Model of 1972 with the 'Golf Ball' proportional spacing

b

4.3 IBM Electric Standard Model Typewriters
a Model C of 1959–65 made in Canada, France, Germany, Holland, U.K., and U.S.A.
b Model D of 1965 onwards

4.4 IBM Electric Executive Model Typewriters with proportional spacing
a Executive AA, first manufactured in 1944, but the main production was from 1946 onwards
b Model C of 1959–65 was produced in Canada, Germany, U.K., and U.S.A.
c Model D of 1965 onwards
d Model 82 of 1972 has a magnetic card and is an Executive 'Golf Ball' with proportional spacing

In 1950 two developments took place that were extremely important to the division's growth. The IBM introduced a completely electric Decimal Tabulation—an invaluable addition to the Company's typewriter's capacity to function more efficiently with statistical material. During the same year, the World Trade Corporation, one of the Company's wholly owned subsidiaries, began the manufacture of IBM typewriters, and has since made a major contribution to the expansion and development of the electric typewriter by opening new markets in countries throughout the free world.

The production of typewriters in pastel colours and with changeable typebars began in 1952. Changeable typebars allowed the typist to replace standard typebars with special symbol typebars when needed. Heavy investments in research during the next two years resulted in the 'Model B' typewriter in 1954. Although it was not a radical departure from the 'Model A', it incorporated many new features including cushioned carriage return, electric ribbon rewind, and multiple copy control. Subsequently, typamatic keys were added to the 'Model B' typewriter. They provided for automatic repeat action on carriage return, spacebar, backspace, hyphen, and underscore keys.

In October 1955 the Electric Typewriter Division was formally established as an autonomous segment of the IBM corporate structure. The Division was completely integrated, developing, manufacturing, and marketing its entire product line.

In 1957 the 'in put out-put' typewriter, which automatically typed computer originated solutions at a rate of ten to twelve characters per second, was introduced. This device is used to feed information into electronic calculators and data processing systems through electronic impulses.

IBM's millionth electric typewriter left the factory in 1958, and in the same year, the '632' electronic typing calculator was introduced which combined the simplicity of electric typewriter operation with the speed and accuracy of electronic calculation. The machine opened a whole new era in office automation to IBM.

Intensive research and development led to the announcement in 1959 of the model 'C' electric typewriter which incorporated such features as personalized touch control and a complete complement of typamatic keys.

In the summer of 1961 the division announced a technological breakthrough which is currently revolutionizing the typewriter industry. That was the introduction of the IBM 'Selectric' typewriter. This new machine printed by means of a single interchangeable sphere-shaped typing element bearing eighty-eight alphabetic characters, numerals, and punctuation symbols. It had no typebars and no movable carriage. The sphere-shaped single printing element, popularly known as 'Golf Ball', was mounted on a small carrier which ran along a cylindric metal bar while typing. Because the writing element moved and not the paper-carrying unit, the need for a conventional carriage was eliminated. For this reason, the 'SELECTRIC' typewriter required less space, vibration was minimized and there was no carriage return jolt. Another important feature of the new machine was the flexibility of type styles offered by the single element typing principle.

IBM manufacture many sophisticated items but in the field of typewriters, they have confined their attention to electric machines and have never made either a standard or portable typewriter.

THE IMPERIAL TYPEWRITER COMPANY, LEICESTER, ENGLAND

In 1902 a young American of Spanish origin named Hidalgo Moya arrived in Leicester with a hand-made model of what he considered to be a revolutionary typewriter. He had acquired a good knowledge of typewriters in America where he had been associated with Hammond.

When he came to England, it took him some time to attract the attention and arouse the interest of manufacturers. Moya was a genius, but not all of his schemes were successful; he made perfectly good violins, but he also built a hot-air balloon, the failure of which was demonstrated when it crashed into a tree!

Eventually however he persuaded a Leicester business man, Mr. J. G. Chattaway (whose daughter he subsequently married) to give him financial assistance for his cherished project of producing a typewriter that would sell for five guineas. He opened a small factory in Garton Street, Leicester, and produced the Moya Model '1', but it cost a good deal more than his estimated five guineas. This machine was not very successful so Moya produced a second typewriter, the Model '2', which was still costly to manufacture although it was an improvement on the Model '1'.

To remain in business, Moya needed more financial assistance, and this he obtained from two local businessmen, W. Evans and J. W. Goddard in 1908. Thus the Imperial Typewriter Company came into existence. The first machine the company produced was the Model 'A' and although this was unconventional in design and appearance, it was based on real scientific and technical knowledge. The machine was widely acclaimed and Leicester's typewriter industry commenced commercial production.

In 1911, the Company moved to new premises in North Evington and the improved manufacturing facilities enabled production of the typewriter to be increased to satisfy the growing demand both at home and abroad. Exports expanded to markets which included France, Russia, and Sweden.

By the beginning of World War I the Model 'B' was introduced, and it differed, not so much in design, as in detail.

Typewriter manufacture ceased entirely during World War I and the plant was turned over to making munitions. Model 'D' was produced in 1919, but did not go into full production until 1921, when further extensions to the factory were made. The Model 'D' was a greatly improved version of the 'down-strike' principle built with better materials and a straight keyboard.

In 1923, the first Imperial portable typewriter was produced; it was a smaller, lighter version of the Model 'D'. In 1927 the Model '50' was introduced. This was the forerunner of the Imperial four-bank machine, offering complete 'interchangeability' features in the form of alternative carriages, platens, and type units. Because of the ease with which foreign keyboards could be fitted to this machine, it opened up many new overseas markets to the Company. Many improvements were made over the years. Production ceased in 1955, but it was the basis of the Standard Imperial as we know it today.

The 'Good Companion' portable was introduced in 1932 and it continued in production first in Leicester, then in Hull, until 1963.

During World War II typewriter production had initially a priority classification and more than 150,000 Imperial machines were supplied for government use.

In 1954 the Company opened their new factory at Hull and announced the Model '66'. Production at the Hull factory was devoted to portables, to enable the Leicester plant to concentrate on the production of the Model '66' and the later standard machines. Imperial commenced the production of an electric typewriter in 1960 with the ELA series.

Immediately after World War II various models of the Imperial 'Good Companion' portable were produced. A radical change in the design of the machine was produced in 1964, and this was known as the 'Messenger'. Production of this machine ceased in 1967. From 1967 onwards, portable machines were imported from Japan and sold under the name of 'Signet' Model '200' and Model 220', also a semi-electric portable, Model '300'. The larger model known as the 'Safari' was at first imported from America and later from Portugal where it is now manufactured.

a

b

4.5 Moya Typewriters (forerunners to the Imperial Typewriters), made by the Moya Typewriter Company, Leicester, England

a Model No. 1 of 1902 is of type sleeve design and belongs to Group 3. It has not completely visible writing and uses rubber type

b Model No. 2 of 1905 is also of type sleeve design belonging to Group 3. However the machine was re-designed to obtain good visibility

The Diamond Jubilee Year for Imperial was 1968. It was also a record year when over 40 per cent of the total production was exported.

In 1969 new records in the export field were reached; a gold medallion award for exports was presented to the Company by the International Export Association, and in 1970, the Queen's Award to Industry was conferred for its outstanding export achievements during the preceding three years. During this three-year period, prior to the Award, the value of Imperial's exports increased by over 40 per cent.

During its history Leicester's typewriter industry has gained an international reputation for producing top quality, reliable typewriters—electric, manual, and portables—and its range of products now include electric and manual adding machines, electronic calculators, and typewriter accessories.

Since becoming associated with Litton Industries Corporation of America, Imperial has expanded its activities on all fronts of the office equipment field, and its present extensive range of products are a direct result of the enormous resources of the parent organization. Imperial and its 4,000 employees are enjoying the fruits of the Leicester-American association in terms of full employment, higher sales and greater earnings. Imperial Typewriters Limited are now assured of a large section of the world's business equipment market which is growing at the rate of 15 per cent per annum.

The latest Imperial model is the Imperial '790', a full-size office electric machine with a ninety-character keyboard, four repeat keys, repeat forward and backspaces, impression and touch controls with four and eight positions respectively, and a wide choice of type styles.

It includes the characteristic Imperial features such as automatic magic margin setters, keyset tabulator, and lateral half-spacer; it has a five-line spacing selector, margin justifier scale, tissue deflector, and a magic monitor that automatically adjusts the impression.

The Imperial '790' with its contoured finger-flow keyboard, has a nylon fabric ribbon. The same machine can be supplied with a film ribbon and in this case it is called the '795'.

There are two carriage widths, 14 in. and 17 in. and the pitches available are 10, 11, 12 and 16.

Table 8

PRODUCTION OF IMPERIAL TYPEWRITERS

Imperial Standard Manual

No. 50

Up to 76,200	1927
76,201	1928
80,251	1929
86,501	1930
95,001	1931
100,001 to 'A' Code	1932
Codes B to D	1933
Codes E to G	1934
Codes H to J	1935
Codes K to N	1936
Codes O to Q	1937
Codes R to T	1938
Codes U to X	1939
Codes Y to Z	1940
Codes ZA to ZC	1941
Code ZD	1942
Code ZE	1943
Code ZF	1944
Code ZG to Z154,999	1945
Z155,000	1946
Z168,200 to Z174,000	1947
Z174,500	1948
Z175,200 to Z175,600	1949
Z175,700	1950
Z176,167	1951
Z176,778	1952
Z177,360	1953
Z177,700	1954
Z177,901 to Z177,999	1955

No. 55

Codes AA0001 to AA1250	1937
Codes AA1251 to AA6110	1938
Codes AA6111 to AB0499	1939
Codes AB0500 to AB3999	1940
Codes AC0001 to AC1699	1941
Codes AC1700 to AC4000	1942
Codes AC4001 to AC4299	1943
Codes AC4300 to AC5199	1944
Codes AC5200 to 262,999	1945
263,000	1946
268,000	1947
284,000	1948
305,930	1949
331,003	1950
349,987	1951
361,461	1952
364,000	1953
364,801	1954
365,301	1955
365,901	1956
366,001	1957
366,201	1958
366,501	1959
366,650	1960
366,900 upwards	1961

Model 58

287,940X	1948
306,380X	1949
331,012X	1950
351,282X	1951
361,454X	1952
364,200X	1953
364,801X	1954
365,301X	1955
365,901X	1956
366,001X	1957
366,201X	1958
366,501X	1959
366,650X to 366,900X	1960

Model 60

6A00006	1949
6A01200	1950
6A10022	1951
6A34371	1952
6A64257	1953
6A73701	1954
6A81801	1955
6A85601 to 6A86500	1956

Model 65

6B00000	1952
6B02800 to 6B17400	1953

cont. page 131

Table 8 contd.

6C00000 to 6C11200	1953
6B17401 to 6B21450	1954
6C11201 to 6C30700	1954
6C30701 to 6C30750	1955

Model 66

6D00000	1954
6D10001	1955
6D46001	1956
6D90001	1957
6E30001	1958
6E68001	1959
6F14001	1960
6F55001	1961
6G06001	1962
6G39701	1963
6G54000	1964
6G59001	1965
6G61301	1966
6G62501 to 6G63226	1967

Model 70

7A01000	1962
7A07500	1963
7A41800	1964
7A80301	1965
7B21301	1966
7B51501	1967
7B89985 to 7B90814	1968

Model 80 (Series I)

7F0001	1968
7F38741	1969
7F83182 to 8F20392	1970

Model 80 (Series II)

8F20393	1970
8F30940 up	1971

Imperial Electric

ELA001	1960
ELA501	1961
ELB101	1962
ELB601	1963
ELC501	1964
ELE001	1965
ELE2001	1966
ELF1000 to ELF1090	1967

660

8,558,054	1967
8,762,635	1968
9,151,681	1969
9,394,485	1970
9,430,420 up	1971

770/775

9,519,180	1970
9,658,954 up	1971

4.6 Early Imperial Typewriter Models

a Model B of 1915 is of the down-strike from the front class and was portable. The typebars strike down into a type guide and onto the vertical centre line of the platen giving visible writing. The keyboard has twenty-eight keys giving eighty-four characters by double shift and is arranged as a sector of a circle

b Model D of 1919 is a later version of Model B and is very similar. However, the keyboard has thirty-two keys giving ninety-six characters by double shift and the keys are arranged in a straight line

4.7 Imperial Standard Manual Typewriters (*see* Table 8)

a Model No. 50 of 1927–55

b Model No. 55 of 1937–61

c Model No. 58 of 1948–59

d Model No. 60 of 1949–56

e Model No. 65 of 1952–5

f Model No. 66 of 1954–64

contd.

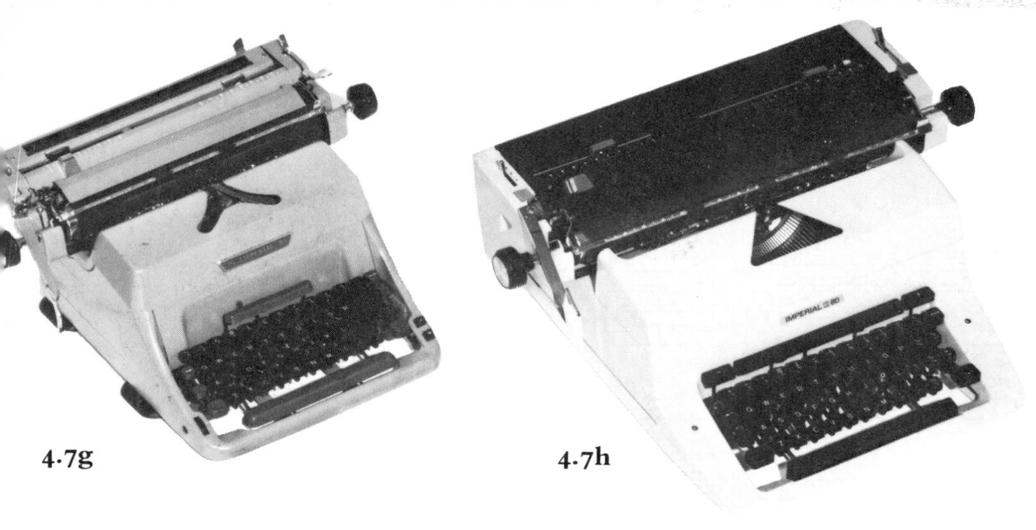

4.7g

4.7h

Imperial Standard Manual Typewriters (contd.)

g Model No. 70 of 1962–8
h Model No. 80 of 1968–70 Series I (1970 onwards Series II)

4.8 Imperial Standard Dual Unit Typewriter of 1955 (Serial No. 365,171). Model 50 This machine had interchangeable specialized and normal keyboards

a

c

b

d

4.9 Imperial Electric Typewriters (*see* Table 8)

a Model of the early ELA-ELF Series 1960–7
b Model No. 660 of 1967–71
c Model No. 777/775 of 1970–1
d Model No. 790 of 1972 which is the latest one

4.10 Imperial Portable Typewriters

n

o

- a Mead Model of about 1928. This was a portable production of the Torpedowerke which eventually became the Good Companion Model
- b Good Companion Model No. 1 of 1932 (Serial No. 1,344)
- c Good Companion Model T of 1942 (Serial No. 2V 895)
- d Good Companion Model No. 3 of 1952 (Serial No. 3AC 454)
- e Good Companion Model No. 4 of 1959 (Serial No. 4X 147)
- f Good Companion Model No. 6 of 1961 (Serial No. 6AR 018T)
- g Good Companion Model No. 7 of 1961 (Serial No. 4X 147)
- h Messenger Model of 1964 (Serial No. 81,691T)
- i Model 1,000 of 1965 (Serial No. SL 17,436,409)
- j Concord Model of 1968 (Serial No. 8,093,575)
- k Signet Model of 1971
- l Model 200 of 1969
- m Model 220 of 1969
- n Safari Model of 1967
- o Model 300 of 1971 which is semi-electric

LUCZNIK: WARSAW, POLAND

The Lucznik typewriter is manufactured in Warsaw in Poland. The model 'FK' was produced between 1933 and 1939 in the National Armaments Factory in Warsaw, and during this period about 15,000 typewriters were produced. During the German invasion in 1939 most of the factory area was damaged and production of typewriters ceased.

In 1970 the company commenced production of the Lucznik which is the Swedish machine 'Facit' made under licence. Most of these manual standard office typewriters are taken by the Eastern European countries, though some are exported to other countries.

a

b

4.11 Lucznik Typewriters
- a Model FK of 1933–9 which was manufactured in Warsaw, Poland
- b Shows view of the current 1972 model made in Poland

4.12 Typewriter assembly in the Lucznik factory, Warsaw, Poland

MESSA (SIEMAG): PORTUGAL

The firm of Siegener Maschinenbau A.G. (SIEMAG) was founded in Siegen in 1890 as an engineering concern manufacturing such things as internal combustion engines, compressors, pistons, and diesel engines.

At the conclusion of World War I the firm was taken over and a new factory was built in Eiserfeld. It was from here that the Siemag typewriter, a four-bank standard office machine, was launched in 1948.

At that time, so soon after World War II, there was a great demand in Germany for typewriters, for there was an inevitable shortage as a result of typewriter production being curtailed or even halted, either by government direction or allied bombing.

The shortage was felt particularly acutely in West Germany, as most of the pre-war German typewriters had been manufactured in what is now known as the Eastern Zone, or German Democratic Republic.

The Company therefore produced both standard and portable machines, but after some years, the plant was transferred to Portugal where a number of models are now manufactured by Messa, S.A.R.L. Mem Martins, Lisbon, and are sold under the name of Messa. The standard machine was also sold in the United Kingdom under the name of 'Sterling' for some time. Messa manufacture for Litton Industries the Royal 'Safari' and the Imperial 'Safari'. In addition they produce a portable based on the 'Patria' design which is marketed in various countries under different names. Consideration is also being given to the production of typewriters in Turkey.

4.13 Messa/Siemag Typewriters which have been sold in England as the Sterling Model
a An early Standard Model of 1948
b Model of 1961 produced in Portugal
c An early Portable Model produced in Portugal

a

b

c

See also page 1

Table 9

PRODUCTION OF STERLING SIEMAG TYPEWRITERS

Sterling Siemag	
M.34	
Up to 15,000	1948—1950
15,001 to 34,000	1951
S.M. 75	
35,000 to 60,000	1952
60,001 up	1953
Standard (Single-Bar Tab)	
De Luxe (8-Key Set Tab)	
100,000 to 108,000	1953
108,001 to 132,000	1954
132,001 to 142,000	1955
142,001 to 146,000	1956
146,001 to 172,000	1957
172,001 to 212,000	1958
214,000 to 231,000	1959
232,000 to 239,000	1960
244,000 to 253,000	1961
253,001 to 262,000	1962
Model 64	
P262,380 to 263,000	1963
263,001 to 268,000	1964
268,001 upwards	1965
Portable	
1-10,000	1955
10,000 to 35,000	1956
Sterling 85	
H6217 upwards	1963

NIPPO PORTABLE TYPEWRITER: JAPAN

This Japanese portable typewriter is manufactured by the Nippo Machine Company Limited and is produced in two designs—the Model 'P200' and Model 'P300'. It is sold in many countries under different names and by various distributors and is illustrated overleaf.

In England it is marketed exclusively by a large group of chemists and known as 'The Nippo'. They are a good quality conventional portable.

The Nippo 'P200': this is a portable typewriter, only 3 in. high, 11 in. in length and width, and weighs only 9½ lb. It can be supplied with type for various languages including English, French, German, and Spanish.

It has a variable line-spacer, a ribbon colour change lever, and a carriage lock lever. The paper support arms are raised automatically by just a touch of its release lever. Half-spacing is easily controlled by moving the backspace key and the space bar, which allows the carriage to move half a space forward. The machine is also equipped with a margin and jammed type-bar release key.

NIPPON: JAPAN

In 1913 the first typewriter with Chinese characters was invented by Mr. Kyota Sugimoto. Figure 4.16 shows the first prototype developed by him.

In 1914 the first machine was marketed by Nihon Shojiki (Nippon Letter Writing Machine Company Limited). In 1916 the patent was applied for in U.S.A. and granted in 1917. In 1917 Nippon Typewriter Company Limited was established and started manufacturing, the typewriters having the moveable keyboard set with Chinese characters, Japanese Hiragana, or Katakana letters.

The firm has now been absorbed by a much larger organization, the Nippon Remington Rand Kaisha Limited. This new company is jointly owned by Mitsui and Sperry Rand Corporation, Remington Rand Division.

Now with more than fifty years experience, they occupy a leading position with advanced models of the kind shown, but the bulk of their output is devoted to electronic accounting machines and other office equipment.

Model '10W-At' (semi automatic with carbon ribbon device)

1. As the Japanese characters (Chinese characters) number more than 10,000, it is necessary for the user of this typewriter to select the most commonly used characters (up to about 3,000) which can be set on to a movable plate (keyboard-type holder).

   If the user needs more than 3,000 characters, the spare movable holder would have to be inserted.
2. Typing is available in both lateral and vertical directions and the pitch is changeable (5–20 pitches) in accordance with the different applications.
3. The size of types (characters) is interchangeable.
4. There is a stencil switch.

Portable Model 'SH-2205' (hand operated)

1. Typing is available only in a lateral direction.
2. The movable holder for this model will hold 2,205 characters for the simpler applications as follows:

   30 numerals for both Arabic and Japanese style.
   75 Katakana letters (Japanese)
   52 alphabet (capital and small letters)
   75 Hiragana letters (Japanese)
   35 symbols
   1,850 Chinese characters (Toyo Kanji)
   28 supplementary Chinese characters
   60 extra spaces for any desired types

.14 The Messa Typewriter of about 1969 was produced for Litton Industries; it was sold in England as the Imperial 'Safari' and in America as the Royal 'Safari'

a

b

4.15 Nippo Portable Typewriters of about 1969 onwards
a Model P200
b Model P300

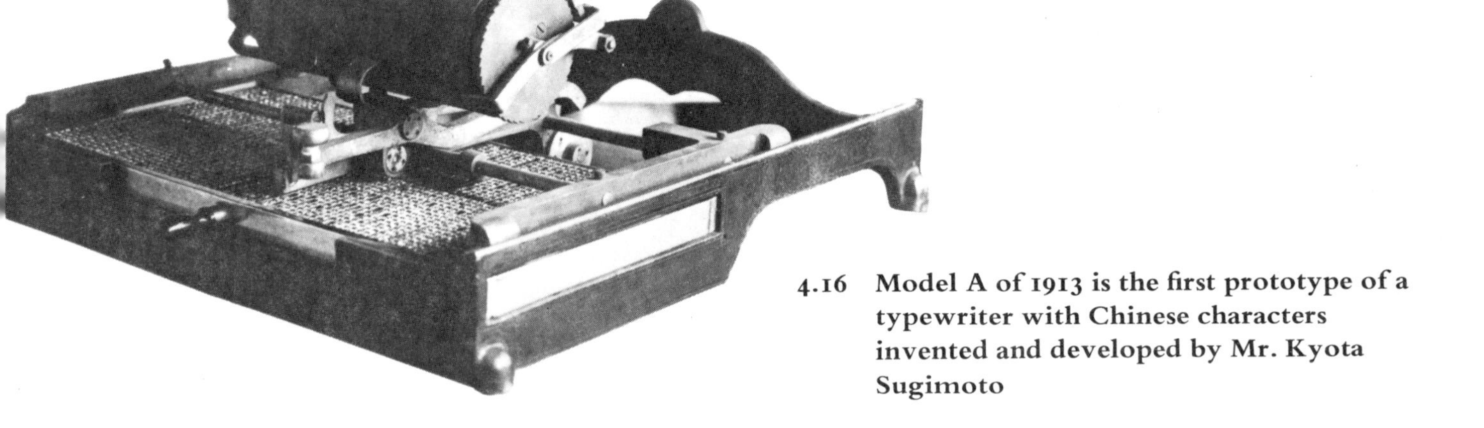

4.16 Model A of 1913 is the first prototype of a typewriter with Chinese characters invented and developed by Mr. Kyota Sugimoto

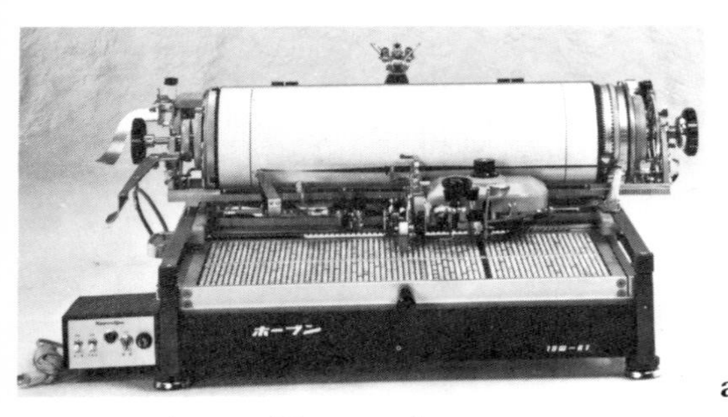

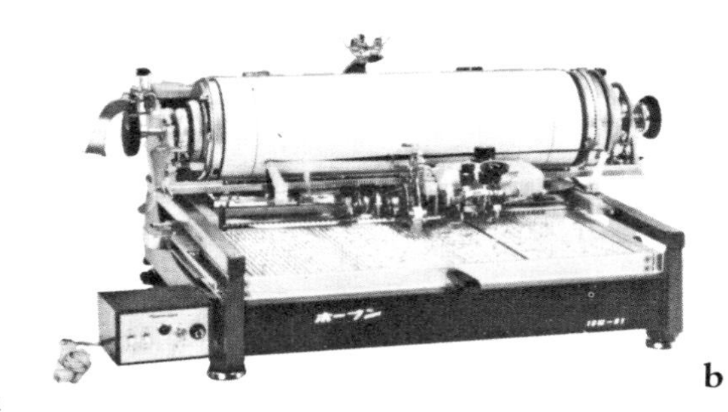

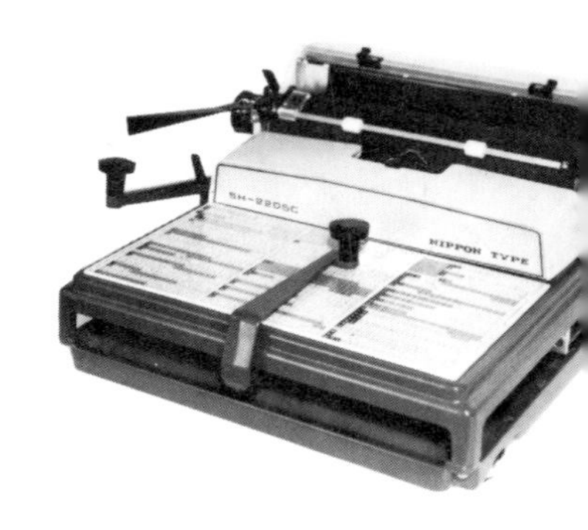

4.17 Nippon Typewriters
a Model 10W-AT of 1972 is an electric machine
b Portable model 10W-AT of 1972
c Model SH-2205C of 1972 is the portable machine

OLIVETTI TYPEWRITERS: ITALY/SPAIN

After taking his degree in electrical engineering, Camillo Olivetti accompanied his tutor, Galileo Ferraris, to the United States and there taught at Stanford University, California, for two years, where he met Thomas Edison. On his return to Italy, Camillo went into industry with his partners, Dino Gatta and Michele Ferrero. Together, they established in 1896 a small electrical precision instrument plant—the first in Italy, at Ivrea, and called it C.G.S. The factory later transferred to Milan and prospered, but Camillo left the Company and returned to Ivrea to design and manufacture typewriters. In 1908 he formed the Ing. C. Olivetti Company, and after a further trip to the United States to study techniques, produced his first typewriter, the 'M1' in 1910. The machine was of original design, with legible characters, a standard keyboard, two-colour ribbon, decimal tabulator, and backspacer. It was exhibited at the World Fair in Turin a year later and the Olivetti Company won an order to supply 100 typewriters to the Italian Navy.

The Company began to expand, and by 1913 was producing twenty-three typewriters a week, and had branch offices in Milan, Rome, Naples, and Genoa.

During World War I the Company produced munitions, rifles, and machine gun parts, and was the only factory in the country producing anti-aircraft fuses and aero-engine magnetos with synchronizing parts to enable aircraft to fire forward through their propellers. During this time the production of typewriters was reduced to a minimum.

After the war, Olivetti returned to manufacturing typewriters, and in 1920 produced the 'M20', which was shown at the Brussels Fair. In the same year Olivetti commenced exporting to Argentina and Holland.

In 1926, the Company diversified and built the Officina Meccanica Olivetti factory for the production of machine tools near the first factory. By this time the number of employees

4.18 Olivetti Typewriters Advertisement.
This is taken from the condensed history of *Typewriter Topics,* **published in 1923, and is the earliest example the author has discovered of an advertisement by Olivetti. It is of interest in that it is in sharp contrast to the futuristic advertising for which Olivetti are now world famous**

had reached 500 and annual production was 8,000 typewriters. The Company was reorganized and able to survive the world depression in 1929, with an annual production of 13,000 typewriters. The first allied foreign Company, S.A. Hispano Olivetti, was founded in Barcelona in that year.

In the middle of the 1920s the Olivetti Company first started welfare schemes for the benefit of their employees to supplement the State Scheme. These were pioneered by Camillo Olivetti, and extended by his son, Adriano. Accommodation, nurseries, medical centres, summer camps for employees and their families, training centres and schools, a library, farm centres, sports and recreational facilities, were provided free. Shorter working hours and

4.19 **Olivetti Standard Typewriters**
a **Model M 1 of 1910**
b **Model M 20 of 1920**
c **Model M 40 of 1930**

Table 10

PRODUCTION OF OLIVETTI TYPEWRITERS

Olivetti Standard Manual	
Standard	
20/6931	1920
M.20	
87179	1933
Model 40	
97060	1931
556000	1947
M.40/3 (All Models)	
400,000 to 550,185	1947
M.44	
1,000,000 upwards	1947
400,000	1948
405,581	1949
433,334	1950
453,500 (grey)	1951
500,000	1952
615,000 upwards	1953
Lexikon	
To 19,999	1954
20,000	1955
40,000	1956
69,000	1957
91,001	1958
117,000 to 125,000	1959
Diaspron 82	
125,001	1959
138,463	1960
164,044	1961
90,020,912	1962
90,050,339	1963
90,076,115	1964
90,099,266	1965
90,126,538	1966
90,153,438 to 90,182,924	1967
Linea 88	
B031,823	1969
B088,010	1970
B179,462	1971

Olivetti Portables	
MP.1	
1790	1932
139706	1950
Studio 42	
504450	1938
694264	1952
Studio 44	
731456	1952
Scribe—Same as Lettera 22.	
Lettera 22	
To 626,232	1954
626,233	1955
636,601	1956
650,001	1957
698,001	1958
724,001	1959
752,419	1960
784,018	1961
91,027,625	1962
91,069,145	1963
91,112,934 to 91,123,040	1964
Lettera 30	
Lettera 31 (Dora)	
4,195,032	1969
5,385,501	1970
Lettera 32	
2,396,736	1965
2,634,090	1966
2,858,298 up	1967
Lettera 33	
Lettera Deluxe	
4,620,363	1969
5,385,501	1970

contd. page 143

Olivetti Electric	
Lexikon 80E	
1,036,680	1956
1,049,636	1957
1,062,065	1958
1,080,122	1959
1,105,116	1960
1,130,422 up	1961
Olivetti 84	
4,107,372	1961
4,161,020	1962
4,213,497	1963
4,228,194 up	1964
Tekne 3	
27,215	1965
107,530	1966
155,803 to 223,093	1967
Tekne 4	
6,501,382	1965
6,504,546	1966
6,519,483 to 6,539,340	1967
Tekne 5 (Editor)	
6,000,351	1965
6,000,870	1966
6,010,785	1967
6,017,973	1968
6,030,000	1969
6,038,000	1970
6,051,000	1971
Praxis 48	
5,180,000	1969
5,227,000	1970
5,283,000	1971
Editor 3	
1,300,000	1970
1,320,000	1971
Editor 3C	
1,000,000	1970
1,010,000	1971

contd. page 143

Table 10 contd.

Olivetti Portables contd.		*Olivetti Electric contd.*	
Studio 44		**Editor 4**	
599,576	1965	500,000	1969
646,487	1966	540,000	1970
1,003,997 up	1967	594,000	1971
Studio 44 (new series)		**Editor 4C**	
		6,600,000	1969
Studio 45		6,615,000	1970
1,210,524	1968	6,638,000	1971
1,321,512	1969		
1,495,079	1970		

longer vacations, child and marriage bonuses, and leave with full pay during pregnancy, were all introduced by the Company as time progressed.

In 1930 the 'M 40' was built, and mass-produced by 1931. The first Olivetti portable, the 'MP 1' followed in 1932, and in 1935 a semi-standard typewriter, the 'Studio 42' was created. By 1939 the Company was exporting 7,400 standard machines and 7,375 portables annually.

During World War II, Olivetti continued to expand and developed their first adding machine, the 'MC 4S Summa' in 1940, using pressure moulding, followed in 1941 by the 'MC 4M Multisumma'.

In 1943 a National Liberation Committee was formed at the factory shortly before the death in hospital of Camillo Olivetti. A year later the Commander of the Partisans, Guglielmo Jervis, was shot and hanged by the Nazis, and in the same year a Nazi plan to destroy the factory was foiled by the factory technicians.

A plaque above the entrance to the factory commemorates those workers who died in the Resistance.

Following the war in 1946, calculating machines and adding machines were produced and established Olivetti internationally in this field.

In 1947 British Olivetti Limited was established in London. They set up a factory in Glasgow and, three years later, the Olivetti Corporation of America was founded.

Landmarks in Olivetti production in post-war years:

1947—The 'M44' replaced the 'M40/3'.
1953—The 'Lexicon 80' typewriter went into production.
1955—The 'Lexicon 80E'. This was Olivetti's first electric typewriter to be followed by the '84E' in 1961.
1957—The 'Graphika', the only porportional spacing manual standard machine was produced.

In 1959 Olivetti took over the Underwood Corporation of America and after a period of rationalization, their machines became synonymous. In some parts of the world they are marketed as Underwood but in other parts they are sold as Olivetti or the Olivetti-Underwood Corporation.

4.20 Olivetti Standard Manual Typewriters
- a Model M40/3 of 1946
- b Model M44 of 1947
- c Lexicon 80 Model of 1954–5. The Olivetti Graphika produced in 1957, has the same external appearance as this machine but it has proportional spacing and is the only manual machine with this feature
- d Diaspron D 82 Model of 1959
- e Linea 88 Model of 1969 is a four-bank machine. The Goffrata Standard Machine is similar to this but is of a different colour with minor internal modifications
- f Linea 98 Model of 1972

4.21 Olivetti Electric Typewriters
- a Lexicon 80E Model of 1955 was the first electric standard
- b Model 84E of 1961 is also a Standard Electric machine

contd.

c Tekne 3 Model of 1965 is a Standard machine and has a two-colour ribbon. The Tekne 4 has a carbon ribbon as has also the Tekne 5 but this machine is fitted with proportional spacing. All the Tekne models have a similar external appearance

d Praxis Model of 1969 is a Compact electric machine

e Editor 4 Model of 1969–70 is a Standard electric machine and has a two-colour ribbon only. The 4C and 5 models have a carbon ribbon only and the 5 also has proportional spacing. All three machines are similar in external appearance. The Editor 3 typewriter has a two-colour ribbon and the 3C has a carbon ribbon only; these two machines are the same in appearance but are slightly lower than the Editor 4 Series

f Lettera 36 Model of 1972 is the first Olivetti electric portable

Olivetti assemble many products in South Africa and South America, and are represented in almost every country. They owe their place in the office equipment world to their founder, and to a far-sighted policy towards product development and staff relationship. They are quite unique in both the advertising and presentation of their products.

All Olivetti typewriters are Group 10 front-strike design.

4.22 Olivetti Portable Typewriters

- a Model MP 1 of 1932 was the first portable machine
- b Studio 42 Model of 1935 was a semi-standard machine
- c Studio 44 Model of 1968 is a desk-top portable machine
- d Studio 45 Model of 1972 is a large portable machine
- e Lettera 22 Model of 1954
- f Lettera 30 and 31 (Dora) Model of 1969
- g Lettera 33 and Lettera De-Luxe Model of 1969 with key tabulator
- h Lettera 32 Model of 1972 with a keyset tabulator

4.23 Shows view of the Electric Typewriter Assembly Shop in the Olivetti Crema Factory, Italy

OPTIMA TYPEWRITERS: GERMANY

The early history of the Optima typewriter is really the early history of the Olympia. The Optima factory, situated in Erfurt in the East Zone of Germany, commenced production of the Olympia Model '8' and the portable machine almost immediately after World War II under the name of Olympia, but the Olympia Directors and many of the staff had moved by then to the West Zone of Germany and opened a factory in Wilhelmshaven where they began producing portables and, later, the Model 'SG1'.

For a short time there was the strange situation of two typewriter factories, one in the Eastern Zone and one in the Western Zone, producing typewriters with the name Olympia. For instance, all the Olympia Model '8s' were produced by the Erfurt factory who now make Optima typewriters.

It was soon established, however, that only the factory in the West producing Olympia typewriters was legally entitled to use the name *Olympia* and, in 1950 Optima began the production of Optima 'M10' followed by the Optima 'M12'.

The current machines are the manual Optima 'M16' and the electric Model 'M100'.

In East Germany, typewriter production has been rationalized and is government controlled. Portables are made by ERIKA; Standard and Electric, by OPTIMA.

The Optima 'M12' is a standard office typewriter produced in 1950. In the same year the

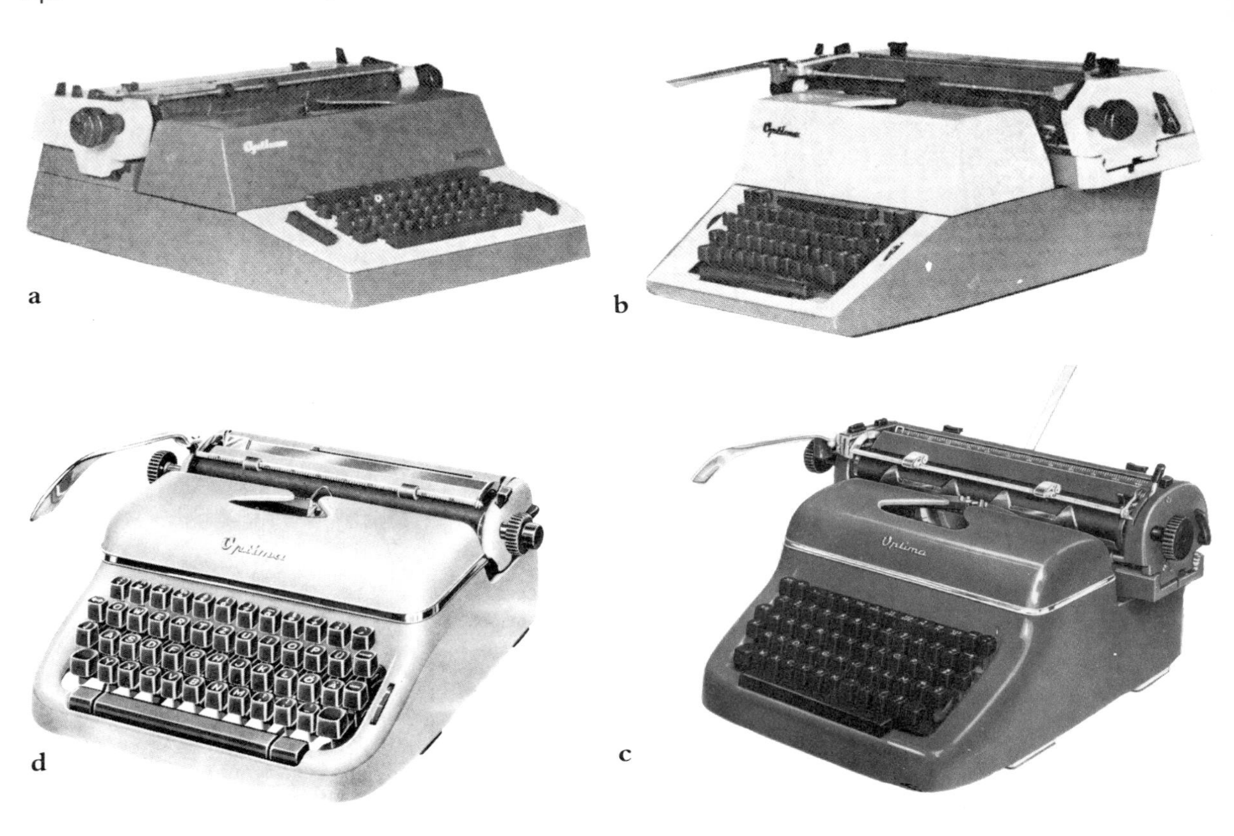

4.24 Optima Typewriters
a Electric model of 1973
b Manual model of 1973
c Model M 12 of 1950
d Elite Model of 1950 which is a portable machine

portable Optima Elite was introduced. It was also supplied with an Arabic keyboard for the oriental market.

Later Optima models include:

Optima 'M16' with a 32 cm carriage.

Optima 'M16/38', like Optima 'M16', but with a 38 cm wide carriage.

Optima 'M16/47', like the 'M16' but with a 47 cm carriage.

CHAPTER 5

History of Typewriter Manufacturers Part III

OLYMPIA INTERNATIONAL: GERMANY

The Olympia typewriter was first produced as the 'MIGNON' in 1903 and gradually improved upon until the 'PLUROTYP' was manufactured in 1933. This was the last production model of the indicator type machine. Altogether 350,000 Mignon typewriters were produced and sold. Originally, the Company was owned by A.E.G. of Germany (Allgemeine Elektrizitäts-Gesellschaft).

This company also produced an 'A.E.G.' typewriter which was a conventional four-bank machine designed in 1912 and ready for production in 1914. However, with the outbreak of war, the plans to produce it had to be deferred. It eventually reached the market in 1921 and was known as the A.E.G. 'Model 3'. The factory was then moved from Berlin to Erfurt. From 1921 until 1933, the Mignon continued to be produced and sold side by side with the A.E.G.

In 1930, the use of the brand name 'OLYMPIA' commenced and a year later, production of a four-bank portable typewriter began. In 1936 the company was renamed 'OLYMPIA'.

In 1939 the first flat portable typewriter came on the market and was called the 'PLANA'.

During World War II the production of typewriters continued, but much of the factory was given over to manufacturing war material. Immediately after the war, the factory in Erfurt, now in the Eastern Zone of Germany, continued to produce the 'Model 8' Olympia and the 'ELITE' portable typewriter, and many of these machines were sold in various parts of the world using the name of Olympia. The true Olympia Company, however, had, by this time, opened a plant to produce typewriters in Wilhelmshaven in the Western Zone of Germany. A dispute as to who should use the name 'Olympia' was eventually resolved in favour of the factory based in Wilhelmshaven. The old factory at Erfurt continued to produce the 'Model 8', followed by other improved standard and portable machines, but these were sold under the name of 'OPTIMA' as they are today. The Optima factory now produces electric and standard machines and not portables.

Olympia went from strength to strength and are today one of the largest manufacturers of typewriters in the world.

By 1953, the Olympia Standard machine, 'S.G.1', was in production and over 7,300

5.1 Mignon Typewriter of 1903–33

a

b

c

d

5.2 Olympia Typewriters
- a Model AEG of 1921–33
- b Standard Model 8 of 1932–50
- c Standard Manual Model SG1 of 1953–65
- d Model SG3 N and L of 1969–73 onwards

3 Olympia Electric Typewriters
a Model SGE 20 of 1958–64
b Standard Model SGE 50 of 1970
c Compact Model SGE 35 of 1970 onwards

people were employed in the Wilhelmshaven factory. By 1956 the number of people employed had increased to 10,400.

In 1957 a new factory was opened in Leer to produce the flat machine. In 1959 'Olympia' joined forces with BRUNSVIGA after introducing their first electric typewriters known as the Olympia 'S.G.E.20/30'. By 1965 a million standard machines had been exported to the U.S.A. and in 1966, the 3 millionth flat portable was produced. By 1968, 2 million standard machines had been sold.

From 1959 onwards, Olympia were also concerned with the production of adding and calculating machines of various kinds and in 1968, began a close co-operation with the Matsushita group (Osaka and Yokohama) for the development of electronic calculators. The Company now produce adding machines in Ireland and assemble typewriters in various parts of the world including Toronto, Chile, Mexico City, and Yugoslavia. Products are exported to 136 different countries through nineteen affiliated companies and 117 main agents. Olympia is a truly international organization with a vast and expanding business potential of which typewriters are now only of secondary importance.

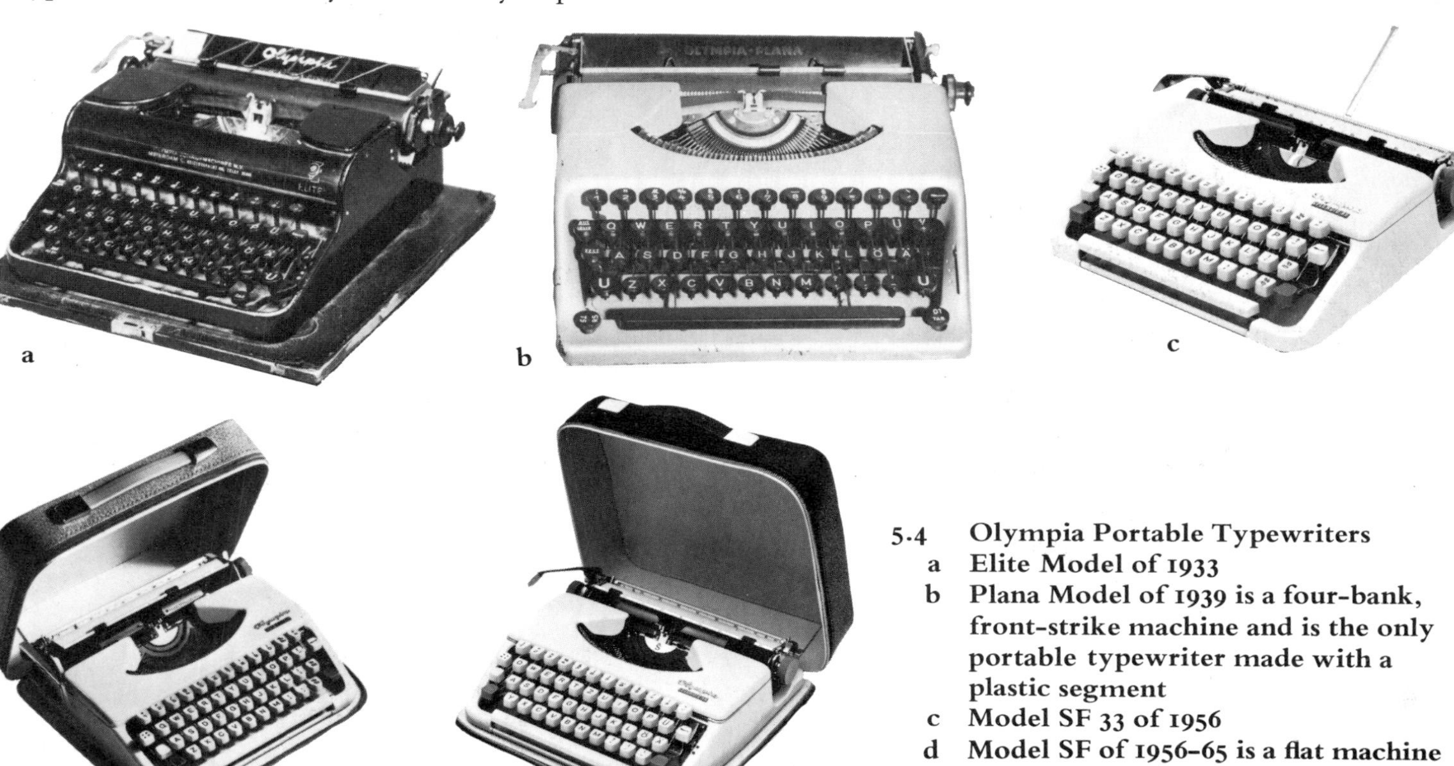

5·4 Olympia Portable Typewriters
a Elite Model of 1933
b Plana Model of 1939 is a four-bank, front-strike machine and is the only portable typewriter made with a plastic segment
c Model SF 33 of 1956
d Model SF of 1956–65 is a flat machine
e Splendid 66 Model of 1966

contd. overleaf

See also Table 11 on page 1

5.4f Model SM8 of 1969

5.4g Model SM9 of 1968 with a 13 in. carriage

5.4h Model SM9 of 1969

REMINGTON: U.S.A.

For all practical purposes the history of the Remington Typewriter Company completely covers the 100 years of typewriter production.

The circumstances that led up to E. Remington & Sons, the gunmakers of Ilion, New York, making a typewriter, have already been told in detail earlier. Undoubtedly, Remington were the first company to put a typewriter on the production line. It was basically a machine invented by Christopher Latham Sholes.

There is much detail available on the company's early history which would be of little interest in a book of this nature. The contract creating the Remington Company as the first typewriter-producing firm was signed on 1 March 1873, and William K. Jenne was given the task of developing the machine for marketing and manufacturing. Actual manufacture began in September 1873 and the first model was shipped from their factory early the following year. It wrote capitals only and used the fundamental features of the inventor's model.

Many arrangements of the Universal keyboard and the carriage return mechanism still remain in present day typewriter construction. The early machines resembled sewing-machines with a stand and foot treadle for return of the carriage, due to the influence of the Remington Sewing Machine Division which manufactured it.

In 1878 the Remington '2' was produced and this solved the very important problems remaining, as it could write in both small and capital letters. In the year 1882 the firm of Wyckoff, Seamans, and Benedict acquired the sole rights for the entire world, and in March 1886 they bought the typewriter business from E. Remington and removed it from its old factory in Ilion.

In 1888 the New York general offices were moved to 327 Broadway where they remained for almost thirty years. At first, only the ground floor and basement were used, but in 1912, all nineteen floors in two buildings from 325 to 331 Broadway, totalling 58,000 sq ft, were occupied.

In 1892 the co-partnership had been turned into a Corporation.

The Remington Company with their familiar slogan, 'to save time is to lengthen life', placed on the market in the summer of 1894 their Model 'No. 6' where the bars were still supported on the old principle, but in 1908 they produced the Model '10', a hanger bar machine, so that the writing was visible. It was a non-stencil switch machine, but this was followed very quickly by another Model '10' with a stencil switch.

In 1897, the Remington Company turned down the rights of the Wagner Writing Machine which later became the Underwood.

In 1920 they produced a further Model '10' with slotted segment and bars.

In 1922, the Company produced the Model '12' which had segment and bars and was enclosed. This was followed by the Model '16' in 1932. Other models were:

5 **Pictures of W. O. Wyckoff (*b.* 1835), C. W. Seamans (1854–1915), and H. H. Benedict (1844–1935), who bought the typewriter business from E. Remington in March 1886**

On Your Travels

WHEREVER YOU GO,
WHEREVER YOU STAY,
TAKE WITH YOU A

Remington

PORTABLE
TYPEWRITER

Full-size standard keyboard, regulation 42 keys, one shift for capitals. Stands 4 inches high with case closed.

The guarantee of REMINGTON world-wide service goes with every machine.

THE SOUNDEST AND MOST COMPACT PORTABLE TYPEWRITER MADE

EASILY STOWED IN THE SMALLEST AVAILABLE SPACE

The cash price of the Remington Portable is £12/10

OR

£2 : 2 : 0

WITH ORDER

THE BALANCE CAN BE PAID BY ELEVEN MONTHLY PAYMENTS OF £1

Call or write for descriptive folder A.S.B.

The Remington Typewriter Co. Ltd.

Head Office: 100 Gracechurch Street, London, E.C.3

Showrooms: 83 Regent St., W.; 4 St. Paul's Churchyard, E.C.4

Branches and Dealers Everywhere

5.6 **An advertisement for a Remington portable from** *Austin Seven Instruction Book* **of about 1922**

Model No. '17', an entirely new construction produced in 1941.
Remington 'KMC' 1946, with keyboard margin control.
The Super Riter Model '18' 1953
Model '20' International GJ, 1953 to 1969 made entirely in the U.K.
Model '24D', is now produced in Italy.

The first Remington electrics were produced in 1925, and there were only 2,500. There was then a large gap. The world was not ready for electrics.

Model '250' Electric was produced in 1953.
Model '300' Electric was produced in 1963.
Model '26' Electric, which is the current machine, was first produced in 1970.

Noiseless machines were either produced by Remington or were the Remington machine made under licence by Underwood. (For the history of the Noiseless Typewriter Co. *see*

contd. page 155

See pages 149–52

Table 11

PRODUCTION OF OLYMPIA TYPEWRITERS

Olympia Standard Manual

4-Bank Model

Serial No.	Year
To 180,000	1932
180,001	1933
188,001	1934
206,001	1935
227,001	1936
243,001	1937
269,001	1938
300,001 to 310,000	1939

S.G. Models

Serial No.	Year
1,001	1953
3,100	1954
27,000	1955
87,947	1956
176,711	1957
287,313	1958
389,968	1959
511,739	1960
646,532	1961
798,651	1962
927,137	1963
1,069,815	1964
1,246,987	1965
1,452,468	1966
1,661,124	1967
1,846,377	1968
2,264,262	1969

Olympia Electric

S.G.E. 20/30

Serial No.	Year
1,001	1958
1,053	1959
2,142	1960
6,713	1961
20,177	1962
24,729 to 26,047	1963

S.G.E. 30 and 30/35

Serial No.	Year
up to 11,979	1967
11,980	1968
30,143	1969

S.G.E. 10/40

Serial No.	Year
1,001	1959
1,063	1960
3,440	1961
7,250 to 13,499	1962

S.G.E. 40

Serial No.	Year
13,500	1963
35,486	1964
77,694	1965
122,905	1966
122,221	1967
243,302	1968
327,019	1969

Olympia Portable

Simplex Model

Serial No.	Year
To 30,000	1934
30,001	1935
80,001	1936
121,001	1937
163,501	1938
200,001 to 224,000	1939

Progress Model

Serial No.	Year
To 10,000	1932
10,001	1933—1934
50,001	1935—1936
80,001	1936—1937
120,001	1938
216,001 to 260,000	1939

Table 11 contd.

Super Model	
To 94,000	1936
94,001	1937
140,001	1938
150,001 to 240,000	1939
Filia Model	
All Serials	1936—1937
Orbis Model	
12,300	1949
SM.I Portable	
12,301	1949
30,001	1950
80,001 to 135,086	1951
SM Portable	
69,390	1950
99,001	1951
206,001	1952
304,001	1953
430,001	1954
590,001	1955
744,181	1956
924,301	1957
1,109,076	1958

Olympia Portable contd.

1,333,916	1959
1,524,843	1960
1,731,685	1961
1,949,310	1962
2,194,889	1963
2,414,049	1964
2,636,381	1965
2,918,569	1966
3,222,681	1967
3,433,392	1968
3,630,904	1969
SF Portable	
1,001	1956
4,863	1957
41,257	1958
155,044	1959
259,318	1960
372,070	1961
483,449	1962
636,791	1963
812,771	1964
982,663	1965
1,152,267	1966
1,345,771	1967
1,542,790	1968
1,708,450	1969

Remington contd.

Chapter 6.) They were based on Kidder's principle, i.e. forward-strike, and the first machine was a cam model three-bank, manufactured in 1909. Then followed:

Model '1' in 1914.
Model '3' in 1917.
Model '5' in 1925.
Model '6' was produced in 1926. It was the first noiseless machine to have four rows of keys.
Models '10' and '11' came between 1934–7.

The de-luxe model with variations of shape, colour, and finish commenced production in 1949 and ceased production in 1969.

In addition to producing their own machines, Remington produced, for many other companies, an enormous variety of standard and portable machines bearing different names. They are still one of the world's largest manufacturers of typewriters. Owing to the high cost of production, however, they now make their standard machine in Italy and Holland and distribute it throughout the world. Compact electric and portable typewriters are manufactured by Brother in Japan and exported to the U.S.A. and all over the world.

They are now a division of Sperry Rand and produce electric calculators, book-keeping machines, and other sophisticated equipment for commerce and industry.

5.7 Early Remington Typewriters
- a Early experimental pre-production Model (Glidden and Sholes)
- b The original production manual Model of about 1874 (Glidden and Sholes)
- c Standard Model No. 2 of 1878
- d Model 10 of 1907 with a non-stencil switch
- e Model 10 manual action
- f Rem-Blick Portable Model prior to 1916

ROYAL TYPEWRITERS: U.S.A.

In the year 1906 Mr. E. B. Hess, a noted inventor, produced the first 'Royal' typewriter incorporating patents which he had secured in 1902. The 'Royal' typewriter had three great assets: it introduced a new idea into typewriter design, it had the services of persevering and determined designers, and lastly, it enjoyed the unwavering financial support of Thomas Fortune Ryan and his associates.

Royal's fundamental principles incorporate over 140 patents which are the property of the 'Royal' Company, and were in the main secured by Mr. Hess.

When 'Royal' introduced Model 1 in 1906, it was designed from the start for the operator, and was a radical departure from previous standards. With its touch which was both light and quick, it required little effort to operate and the accelerating typebar action contributed to faster typing. It was a front-stroke machine with segment, bars, and visible writing (the long carriage version was known as the Model 3).

Through its loyal financial backing, the Company was able to weather the first perilous and uncertain years. The small single storey factory in Brooklyn spread to four floors, and finally in 1908 was moved to Hartford in Connecticut where a large source of skilled labour could be found.

In 1911, the 'Royal' Model 5 was introduced (the long carriage versions were known as 6, 7, and 8). This new machine offered many mechanical improvements. It had a two-colour ribbon and incorporated the first paper bail in any typewriter.

Model 10 was introduced in 1914, and improved upon in about 1928. The 'Q' or 'quiet'

See pages 152–6

a b c d

5.8 Remington Typewriters

a Standard Model EJ Series of 1950–1
b Super Riter Model SJ Series up to 1960
c Standard International Model GJ Series of 1961–9
d Standard Manual Model 24 of 1968–71

See pages 152–6

5.9 Remington Electric Typewriters
a Model of 1925 with an electric action manual carriage return. Only 2,500 of the machines were made and the project was abandoned due to lack of interest
b Model 250 of 1955
c Aristocrat Model of 1960–1
d Model 300 of 1961–2
e Model 25 of 1963–71

5.10 Early Royal Typewriters (All Royal Typewriters are Group 10 four-bank machines)
a Standard Manual Model 1 with a 10 in. carriage. (Model 3 has a longer carriage of 14 in. or 18 in. All three models have a single colour ribbon and the Serial Nos. begin at 1,000 in 1906 to 5,800 in 1911. Model 5 looks the same and has a 10 in. carriage with Models 6 having a 14 in., 7 an 18 in., and 8 a longer one; all have a two-colour ribbon and the numbering jumps to 8,500 in 1911 to 175,000 in 1914.)
b Standard Model 10 of 1923 with a double panel (Serial No. X 575,028)

contd. opposite

c Standard Model 10 of 1929 with a single panel (Serial No. X 1,318,002)
d Model H of 1935. This changed to Model KHM in 1936 and had a keyset tabulator
e Magic Margin Standard Model of 1938 (Serial No. KMM 2,412,403)

model was introduced a little later, but this was unsuccessful and was soon withdrawn. Shift freedom or new-style drop segment began, firstly with the long carriage machines and later over the entire range.

In 1935 the 'H' model came on the market. This was the same as the Model 10, but had a re-styling of the top plate. This was followed in 1936 by the 'KHM' which had introduced for the first time a keyset tabulator.

In 1939 'Royal' introduced the magic margin, and a new style top covered with improved touch control. At that time the company held the coveted position of the world's number one typewriter manufacturer in volume, sales, and outstanding features.

After World War II, 'Royal's' progress continued. Firstly with the 'HH' machine introduced in 1954, and then the Model 'FP', three years later. These were followed by Model 101 in 1963, the 'MC' series or 'Empress' introduced in 1965, and the Model 440 in 1966.

'Royal' produced an electric typewriter in 1950. In design it was deliberately kept as near as possible to the 'Royal' Standard.

Between 1954 and 1956 the 'RE' model, designed as a co-ordinated element, was developed. As the popularity of the electric typewriter increased, additional models were introduced; the first of these was the 'HE' followed by the 'EB'.

In 1962, production of the 'Electress' commenced. It is worthy of special mention, because it was the first electric with an action created by a mathematical formula produced on a computer. This computer design resulted in the elimination of one-third of the operating parts of the conventional electric, and made possible substantial cost reduction.

The de-luxe version of this machine, with built-in carbon ribbon, was known as the 'Emperor'.

In 1966, 'Royal' announced the Royal 660 and also the 550, which was a model specially designed for schools.

Since 1950, Royal have introduced many variations of type styles and colour, though the tendency today is to limit the number of these purely for economic reasons.

The company introduced portable machines in 1926, and modern Royal portables are a variation of the basic design; over 6 million of these have been sold. Today Royal market both manual and electric versions.

For economic reasons many Royal portables (identical with the 'Imperial' typewriter range) are manufactured in Japan and Portugal.

The Visible Writing Machine Company no longer distribute Royal typewriters in the U.K.

Table 12

PRODUCTION OF ROYAL TYPEWRITERS

Royal Portable

Serial number	Date
To 117,000	1927—1928
117,001	1929—1930
230,001	1931—1932
290,001	1933—1934
371,001	1935
476,301	1936
575,001	1937
695,001	1938
740,001	1939
923,128 to 938,350	1940
1,183,274 to 1,236,115	1946
1,236,938 to 1,386,136	1947

Safari

Serial number	Date
5,584,967	1964
5,958,589	1965
6,474,796 to 7,022,435	1966

Signet

Serial number	Date
Incomplete Portable	1932—1933

Junior

Serial number	Date
To 45,000	1935—1936
45,001	1936
62,001 and over	1937

'UR' Model

Serial number	Date
71,000	1937
80,001 and over	1938

'D' Model

Serial number	Date
To 100,000	1939
114,526 to 120,935	1940

Royalite and Royaluxe 400, 425 and 450

Serial number	Date
356,38	1955
3,123,513	1956
3,384,902	1957
3,745,091	1958
3,970,557	1959
4,272,034	1960
4,544,659	1961
4,927,715	1962
5,292,116 to 5,733,104	1963
5,743,811 to 6,095,238	1964
6,098,916	1965
6,580,453 to 6,948,198	1966

Royaluxe 300 and 325

Serial number	Date
104,311	1959
107,687 to 122,040	1960

Diana Portable

Serial number	Date
13,365	1953
22,020	1954
32,144	1955
37,141	1956
43,336	1957
64,701	1958
81,475 and over	1959

Royal Electric

(approximate serial number starting and finishing dates)

Serial number	Date
5,827,414 to (RE Model)	1956
6,899,604 (RE Model)	1960
7,003,601 (HE Model)	1960
7,580,644 (HE Model)	1963

G.A. Model

Serial number	Date
7,663,992	1963
7,820,050	1964
8,026,200	1965
8,215,835 to 8,529,725	1966

E.B. Model

Serial number	Date
7,672,525	1963
7,739,084	1964
8,003,200 to 8,322,305	1965

Royal Standard Manual

Serial number	Date
Nos. 1 and 3	1906—1911
Nos. 5, 6, 7 and 8	1912—1918

contd. page 161

Table 12 contd.

Royal Standard Manual contd.

No. 10

Serial No.	Year
175,001	1914—1916
281,001	1917—1923
750,001	1924—1926
*1,000,001	1927—1929
*1,300,001	1929—1931
*1,470,001	1932
1,520,001	1933
1,615,001	1934
1,670,001	1934
1,679,001	1935
1,829,001	1936
1,990,001	1937
2,170,001 to 2,250,000	1938

Prefix 'Y' 100,000 up treat as 750,001 to 1,000,000.
Prefix 'SY' and 'CSY' 100,000 up treat as 1,470,001 to 1,550,000.
Prefix 'HY' treat as 1,670,000.
Prefix 'KHY' treat as 1,890,000.
Double Glass Panels began at 212,000.
Single Glass Panels began about 745,000.
Variable Line Spacer began at 282,000.
*New Style Drop Segment began as under, 10in. 1,470,000; 14in. 1,330,000; 18in. and wider 1,200,000.

Model MM

Serial No.	Year
2,250,001	1938
2,273,001	1939
2,420,239 to 2,564,583	1940
3,100,452 to 3,197,214	1946
3,215,324 to 3,409,889	1947

Model HH

Serial No.	Year
5,100,000	1954
5,456,674	1955
5,728,700 to 5,989,001	1956

Model F.P.

Serial No.	Year
6,159,809	1957
6,361,664	1958
6,605,188	1959
6,867,976	1960
7,069,762	1961
7,292,351 to 7,470,959	1962

Model 101

Serial No.	Year
7,471,771	1963
7,628,767	1964
7,902,159 to 7,933,672	1965

M.C. Empress

Serial No.	Year
7,942,375	1965
8,216,377 to 8,289,999	1966

440

Serial No.	Year
8,390,000 to 8,550,000	1966/8

All their assets and distribution organization have been taken over by the Imperial Typewriter Company of Leicester, a Division of Litton Industries, to avoid duplication. This policy is varied in different parts of the world.

Illustrations of Royal machines continue at pages 164–6, which see.

S.C.M.: THE SMITH-CORONA STORY: U.S.A.

The history of the Smith-Corona really begins with Alexander Brown who was a talented designer whose inventions had included guns, the Dunlop tyre, and steering gear for cars.

In 1876 Alexander Brown had seen a typewriter at an exhibition in Philadelphia and was convinced he could build a better one himself. Soon after this, he was visited by two of the four Smith brothers, Lyman and Wilbert, who wanted his help in re-designing a gun being manufactured in the Smith factory at Syracuse, U.S.A.

Table 13

PRODUCTION OF SMITH CORONA TYPEWRITERS

Standard Machines

Nos. 1 and 2 Style (Models 1-2, 10″; 3, 14″; and 6, 20″): Approx. 1911. No Type Guide.

No. 5 Style (Models 3, 14″; 5, 10″; and 6, 20″). Type Guide and 4″ Tabulator Bar. 1911–1920.

No. 8 Style (Model 3, 12″, 14″ and 18″; Model 6, 20″ and 26″; Model 8, 10″). 5 Separate Tab. Keys.

Down Shifts commenced as follows:

10″, 440,000; 12″, 4,000; 14″, 56,000 (18″ were all Down Shift); 20″ and 26″, 22,500.

Commencing 1928 until the introduction of the 'Super Speed' all L. C. Smiths, irrespective of carriage lengths, were No. 8 Models, and serial numbered consecutively.

No. 3 (8 Style) 12 in. Carriage

1,300 to 10,528	1921—1923
10,529 to 701,500	1924—1926
701,501 (see No. 8 numbers)	

No. 3 (8 Style) 14 in. Carriage

26,400 to 33,450	1915—1917
33,451 to 55,950	1918—1922
55,951 to 600,000	1923—1926

No. 3 (8 Style) 18 in. Carriage

Below 10,000 (All Down Shifts)	1925

No. 6 (8 Style) 20 in. *and* 26 in. Carriages

7,930 to 18,525	1915—1922
18,526 to 600,000	1923—1926

No. 7 (8 Style) 7 in. to 10 in. Carriages

To 17,242	1915
17,243 to 17,490	1916
17,491 to 17,750	1917

No. 7 (8 Style) *contd.*

17,751 to 18,003	1918
18,004 to 18,500	1919
18,501 to 19,100	1920
19,101 to 19,500	1921
19,501 to 20,280	1922
20,281 to 20,799	1923
20,800 to 21,060	1924
21,061 to 21,582	1925—1926
21,583 to 21,854	1927
21,855 to 22,168	1928
22,169 to 22,629	1929
22,630 to 22,857	1930
22,858 to 22,952	1931
22,953 to 22,963	1932
22,964 to 25,011	1933
25,012 to 25,099	1934
25,100 to 25,138	1935
25,139 to 25,202	1936

No. 8 Standard and Silent

To 317,000	1915—1917
317,001 to 443,000	1918—1922
443,001 to 560,620	1923—1925
560,621 to 661,900	1926
661,901 to 756,420	1927
756,421 to 861,323	1928
861,324 to 939,548	1929
939,549 to 999,229	1930
999,230 to 1,036,436	1931
1,036,437 to 1,049,999	1932
1,050,000 to 1,087,914	1933
1,087,915 to 1,146,469	1934
1,146,470 to 1,209,168	1935
1,209,169 to 1,285,375	1936

No. 8 and Super Speed

1,250,001 to 1,397,858	1937
1,397,859 to 1,457,164	1938

No. 1

1,457,165 to 1,525,092	1939
1,525,093 to 1,616,501	1940
1,616,502 to 1,726,786	1941
1,726,787 to 1,788,559	1942
1,788,560 to 1,793,283	1943
1,793,284 to 1,850,019	1944
1,850,020 to 1,917,387	1945
1,917,388 to 2,002,559	1946

No. 1 *contd.*

2,002,560 to 2,108,698	1947
2,108,699 to 2,193,186	1948
2,193,187 to 2,281,855	1949
2,281,856 to 2,358,620	1950
2,358,621 to 3,103,000	1951
3,103,001 to 3,200,000	1952

Model 88

4,019,932 to 4,025,000	1953
4,025,001 to 4,098,500	1954
4,098,501 to 4,175,000	1955

Model 62

5,050,000 to 5,097,100	1958
5,097,101 to 5,170,754	1959
5,170,755 to 5,268,535	1960
5,268,536 to 5,310,072	1961
5,310,073 up	1962

Model 72

5,314,336 to 5,316,494	1962
5,316,495 to 5,512,300	1963
5,512,301 up	1964

Model 75

75/E 6,151,101 to 6,159,452	1964
75E/ 6,159,453 up	1965

Model 2E (Electric Standard)

5,038,009 to 5,151,831	1958

3E Electric Standard

3E 5,007,213 to 5,027,438	1958
3E 5,027,439 to 5,097,540	1959
3E 5,097,541 to 5,268,042	1960
3E 5,268,043 to 5,268,100	1961
3E 5,268,101 up	1962

Model 400 (4G)

4E 6,000,520 to 6,024,347	1962
4E 6,024,348 to 6,102,630	1963
4E 6,102,631 to 6,114,372	1964
4E up	1965

Model 410

4E 6,132,235 up	1965

Table 13 contd.

Standard Machines *contd.*

Electric 'Junior' 10 in.

Serial numbers	Year
5TE 132,611 to 134,755	1958
5TE 134,756 to 134,837	1959
5TE 134,838 to 135,353	1960
5TE 135,354 to 186,850	1961
5TE 186,851 up	1962

Electric 'Junior' 12 in.

Serial numbers	Year
5LE 101,912 to 113,134	1959
5LE 113,135 to 122,725	1960

Portables

Corona

3, Special, and Plus

Serial numbers	Year
To 33,000	1906—1913
33,001 to 150,000	1914—1918
150,001 to 500,000	1919—1922
500,001 to 600,000	1923—1926
600,001 to 650,000	1927—1931
650,001 to 674,000	1932—1934
674,001 to 683,000	1935
683,001 to 686,000	1936
686,001 to 689,600	1937
689,601 to 695,000	1938
695,001 to 700,000	1939

Corona No. 3. Weight (in case) $9\frac{1}{2}$ lb. Paper 9″. Line $7\frac{4}{5}$″. Dimensions (in case), length $11\frac{1}{4}$″, width $9\frac{3}{4}$″, height $4\frac{1}{2}$″.

Corona Special & Plus. Weight (in case) 10 lb. Paper $9\frac{3}{4}$″. Line $8\frac{3}{10}$″. Dimensions (in case), length 12″, width 10″, height $4\frac{1}{2}$″.

Corona Four

Serial numbers	Year
First Letter E and H	1924–1925
First Letter K, L and M	1926–1928
First Letter P and A	1929–1930
Last Letters AO and BO	1931–1932
Last Letters CO, DO and EO	1933–1934
Last Letters HO	1935
Last Letters KO	1936
Last Letters LO	1937
Last Letters MO	1938
Last Letters PO	1939

Weight (in case) 12 lb. Paper 10″, Line $8\frac{3}{10}$″. Dimension (in case), length 13″, width 12″, height $4\frac{7}{8}$″.

Skyriter

'2' Series

Serial numbers	Year
To 2Y 8550	1949
2Y 8550 to 50571	1950
2Y 50572 to 92427	1951
2Y 92428 to 135042	1952
2Y 135043 to 191002	1953
2Y 191003 to 239329	1954
2Y 239330 to 290373	1955
2Y 290374 to 334611	1956
2Y 337287 upwards	1958

'3' Series

Serial numbers	Year
To 3Y 46383	1956
3Y 46383 to 106810	1957
3Y 106811 to 118237	1958
3Y 118238 to 197,843	1959
3Y 197,844 up	1960

Corona Standard

Serial numbers	Year
To 1C 135,000	1937
To 2C 198,000	1938
To 2C 202,000	1939

Alexander Brown not only helped them in re-designing their gun, but succeeded in interesting the Smith brothers in financing the production of a new and improved typewriter. As a result of this, the double keyboard Smith-Premier typewriter was produced. The machine was so favourably received, that, within a year, Lyman and Wilbert discontinued making guns entirely and, with two other brothers, founded the SMITH-PREMIER TYPEWRITER COMPANY to concentrate entirely on typewriters.

By 1890, the typewriter was no longer regarded as a novelty but had become established as an instrument of business. Unfortunately, there were about thirty different manufacturers of typewriters at this time, and they all had a different idea for placing the letters of the alphabet on the keyboard. This created great confusion in the minds of the public as well as within the industry. The keyboard was not standardized until about 1895 when the basic Sholes' keyboard was adopted. Also, many manufacturers had double keyboard machines, without shift. Others had single keyboards with shift. Some produced only capital letters, whilst others printed upper- and lowercase characters.

See pages 157–61

a b c d e f

5.11 Royal Standard Manual Typewriters

a Model HH of 1954
b Model FP of 1957
c Model 101 of 1963
d Empress Model of 1965
e Model 440 of 1966
f Royal 470 Standard Model of 1972

5.12a

5.12b

5.12c

5.12d

5.12e

5.12f

g-j are Compact and Portable electric machines

g

h

i

j

5.12 Royal Electric Typewriters

- a Model of 1956 H.E. Series
- b Electress Model of 1962
- c Emperor Model of 1962 with ordinary and carbon ribbons
- d Model 660 of 1966
- e Model 550 of 1966
- f Standard Model 560 of 1972
- g Standard Model 970 of 1972 which is produced in Germany
- h Model 1300 of 1972 in the Award Series
- i Apollo Model 10 of 1972 which is a Portable Machine
- j Model 1200 of 1972 which is a Portable Machine

In 1893, Smith-Premier merged with six other leading manufacturers, 'Monarch', 'Brooks', 'Remington', 'Caligraph', 'Densmore', and 'Yost' thus forming the Union Typewriter Company of America. The Smiths were executives in this organization and continued to manufacture the Smith-Premier at their Syracuse Works. During the next ten years, the Smith-Premier became the most popular of the 'Blind' Writing Machines. It had a 'double' keyboard and was promoted as having 'a key for every character'.

The Smiths' place as pioneering leaders in the typewriter industry was summed up in their slogan 'The pen is mightier than the sword, but the Smith-Premier Typewriter bends them both'!

In 1896, controversy arose over a revolutionary new typewriter which used a 'visible' writing principle, whereby the operator could see the material as it was being typed. Previously it had been impossible to see what was being written without lifting the carriage.

The Smith brothers were quick to see the advantages of this method and tried to convince their associates in the Union Typewriter Company to adopt it. They met with little success, so the Smiths resigned and set up their own business producing 'Visible' Writing Machines. On 27 January 1903, the Smith brothers formed the 'L. C. Smith & Brothers Typewriter Company of Syracuse'. Lyman was elected first President, Wilbert took charge of manufacturing, Monroe handled sales, and Hurlbut became treasurer.

In order to produce the quality and quantity of typewriters planned, they left their old gun works and built a new modern plant. In December of that year, the plant was completed. When they moved, the Smiths took with them Carl Gabrielson, a top engineer of the Union Typewriter Company, who was to take over the designing of the new machine. This combined team produced a visible single keyboard model which was completed in late 1904.

See pages 157–61

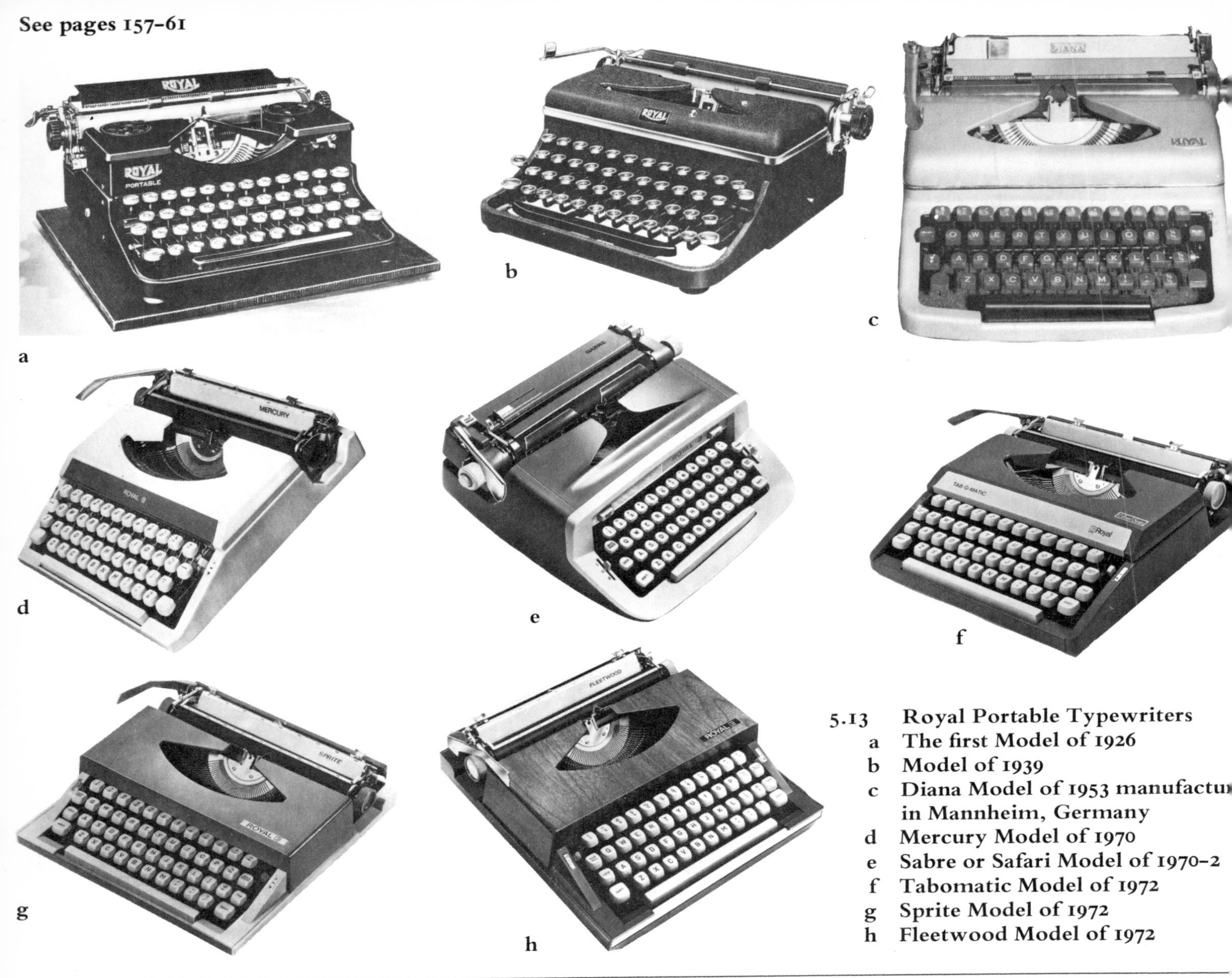

5.13 Royal Portable Typewriters
- a The first Model of 1926
- b Model of 1939
- c Diana Model of 1953 manufactu in Mannheim, Germany
- d Mercury Model of 1970
- e Sabre or Safari Model of 1970–2
- f Tabomatic Model of 1972
- g Sprite Model of 1972
- h Fleetwood Model of 1972

The following features contained in this model were to become standard in future office machines:

1. Segment shift, whereby in shifting for capital letters, the light segment carrying the typebars was moved instead of the carriage.
2. Inbuilt tabulator enabling an operator to move the carriage quickly and accurately to any pre-determined position.
3. Stencil cut-out.
4. Interchangeable platen making it possible to handle regular correspondence and heavy manifolding on the same machine by changing from soft to hard platen.
5. Two-colour ribbon.

Mention should now be made of a separate venture. An American Senator, Ben Conger, had been intrigued by a folding, visible typewriter, invented by a Mr. Frank Rose and his son. Conger was so impressed that he contacted two of his friends and together they bought out the patents and full manufacturing rights of the Rose Typewriter Company. They formed the Standard Typewriter Company and began work on the standard folding typewriter which appeared in 1907. It was light, weighing only $5\frac{1}{2}$ lb and was an immediate success. It is reported that one of these machines was owned by President Theodore Roosevelt who took it with him on safari in 1910.

In 1914, the name of the Company was changed from 'The Standard Typewriter Company' to 'Corona Typewriter Company'.

Some time during 1924, the world famous Dempsey-Firpo fight took place when Firpo knocked Jack Dempsey through the ropes and he landed on a Corona portable typewriter! Standing on the Corona he climbed back into the ring. Miraculously the typewriter still worked. Corona publicity men highlighted the event by saying 'Dempsey knocked out Firpo, but he could not knock out Corona'!

In January 1926, L. C. Smith merged with the Corona Typewriter Company (who were the leading manufacturers of portable typewriters) and the two machines complemented each other perfectly.

During the depression, trade declined and late in 1932, Wilbert Smith retired and H. W. Smith took over his duties as President, electing Elwyn Smith as his assistant.

In 1934, the company took over the Portable Adding Machine Company, introduced a quieter typing 'Silent L. C. Smith' office machine and the 'Silent' and 'Sterling' portables.

In 1935, they introduced the 'Standard' portable known as the 'Clipper'. In 1938 a light-weight 9 lb portable called the 'Zephyr' was introduced, and by 1940, was on a sound financial footing.

During World War II, the factory produced percussion primers for bombs. Its sale force assisted the government in purchasing much-needed used typewriters.

In October 1942 all major companies were told by the American government to halt the

c

d

b

e

f

5.14 L. C. Smith Typewriters
- **a Standard Model 2 which was the first Model of 1903**
- **b Model of 1904 with a front-strike, no type guide, but an upshift ball-bearing action**
- **c Model 1 of 1914 which was the second Model**
- **d Standard Model 7 of about 1920**
- **e Standard Model Early 8 left-hand machine of about 1929 onwards**
- **f Standard Super-Speed Model of 1937 onwards**

5.15 Smith Corona Standard Manual Typewriters
- a Standard Manual Model 62/72 of 1958–
- b Standard De-Luxe Model of 1964–5

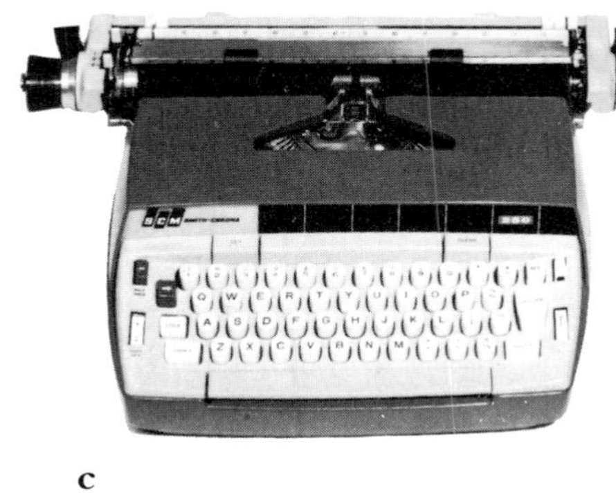

5.16 Smith Corona Electric Typewriters
- a Standard Model 400/410 of 1962–5
- b Model 410 of 1965
- c Compact Model 250 of 1963–5
- d Secretarial Compact Model of 1970 onwards

5.17 Smith Corona Portable Typewriters
- a Corona Standard Folding Model of about 1907
- b Corona Model of 1912 onwards is a three-bank machine
- c Special Telegraphy Machine of 1943 for the U.K. Government only

contd. opposite

5.17d Model of 5 and 6Y Series of 1961 onwards made in the U.K. (formerly called Empire Corona)

production of typewriters. The company could only continue to exist during this period by making rifles, parts for bombs, pistols, machine guns, naval guns, torpedoes, and various other instruments. It had a production of approximately 23,000 Springfield rifles per month.

In November 1943 the Army told H. W. Smith that they had amassed a sufficient number of rifles and now needed typewriters, so he began a four-month reconversion period, but the Company was allowed to finish the rifles they had already begun. Fortunately, Hurlbut had retained all his equipment for making typewriters believing that he would be able to return to it eventually.

In 1946 the company changed its name to 'Smith-Corona' and was manufacturing other machines besides typewriters. In 1953 they changed it again to 'Smith-Corona Inc.'.

They kept abreast of developments during the boom after the war and ten years of research culminated in the production of an electric office typewriter in 1955. It had the following exclusive features:

1. The keyboard was scientifically sloped so that it fitted natural finger movements.
2. There were more controls in the keyboard area than on any other electric typewriter.
3. A signal light visible from anywhere in a room indicated when the machine was on.
4. Automatic repeat actions were available on any or all keys.
5. A highly perfected roll-free printing action was incorporated.

In September 1958 they merged with Marchant Calculators forming 'Smith-Corona Marchant Inc.', thus building an aggressive, diversified Corporation. This Corporation has increased the sales both at home and abroad and is now a well established and well known company throughout the world, with over 10,000 men in its organization. It has more than 7,500 shareholders, many of whom are company employees.

They are now a division of Kleinschmidt and no longer manufacture manual and electric standard office typewriters. They do, however, produce a large quantity of portable machines and compact electric typewriters as well as many other products.

THE TRIUMPH TYPEWRITER: GERMANY

The forerunner of the 'Triumph' typewriter was the 'Norica' which was produced in 1907.

By 1910 it had been re-designed and was one of Germany's best known machines. It was produced with six different carriage widths and all manner of keyboards. It was originally

contd. page 172

5.18 The Norica Typewriter of 1907 is believed to have been manufactured by Triumph A.G., Nuremberg, Germany. It had four rows of keys, belonged to Group 8, and was of semi front-strike design with visible writing; very few of these machines were sold

Table 14

PRODUCTION OF TRIUMPH-ADLER-VERTRIEB TYPEWRITERS

Triumph (Grundig Triumph)
Standard

Serial numbers	Year
750,000 to 950,000	pre 1956
950,001 to 1,000,000	1956
1,000,001 to 1,079,500	1957
1,079,501 to 1,370,000	1958
1,370,001 to 1,389,000	1959
1,389,001 to 1,461,000	1960
1,461,001 to 1,800,000	1961
1,800,001 to 2,000,000 and 5,000,000 to 5,020,000	1962
5,020,001 to 5,136,000	1963
5,136,001 to 5,150,000 and	1962
8,120,001 to 8,130,000	1964
8,130,001 upwards	1965

Electrics

Serial numbers	Year
1,300,001 to 1,301,739	1957
1,301,740 to 1,305,442	1958
1,305,443 to 1,309,056	1959
1,309,057 to 1,312,680	1960
1,312,681 to 1,315,000 and also 1,340,001 to 1,340,919 and also 7,200,044 to 7,200,979	1961
7,200,044 to 7,200,979	1962
7,200,980 to 7,204,533	1962
7,204,534 to 7,207,847	1963
7,207,848 to 7,209,172	1964

Electric Mantura 'S'

Serial numbers	Year
1,315,001 to 1,316,901	1958
1,316,902 to 1,320,647	1959
1,320,648 to 1,324,047	1960
1,324,048 to 1,326,700 and also	1961
1,350,001 to 1,350,951	
1,350,952 to 1,351,000 and also	1962
7,210,000 to 7,211,895	
7,211,896 to 7,213,295	1963
7,213,296 to 7,213,856	1964

Electric 20

Serial numbers	Year
1,725,001 to 1,726,930 and also	1961
7,000,000 to 7,002,686	
7,002,687 to 7,006,079	1962

Electric 31

Serial numbers	Year
7,010,000 to 7,012,505	1962
7,012,506 to 7,019,999 and also	1963
7,040,000 to 7,040,379	
7,040,380 to 7,041,999	1964

Electric 51

Serial numbers	Year
7,030,000 to 7,030,804	1963
7,030,805 to 7,031,999	1964

Electric 31/51

Serial numbers	Year
7,301,000 to 7,344,174	1964
7,344,175 up	1965

Portables
Personal Portable

Serial numbers	Year
4,204,000 to 4,272,000	1961
4,272,001 to 4,300,000	1962
4,300,001 to 4,369,000	1963
4,369,001 to 4,500,000	1964
4,500,001 upwards	1965

Perfect Portable

Serial numbers	Year
1,282,000 to 1,300,000 and 2,000,000 to 2,050,000	1961
3,045,000 to 3,050,000	1962

b

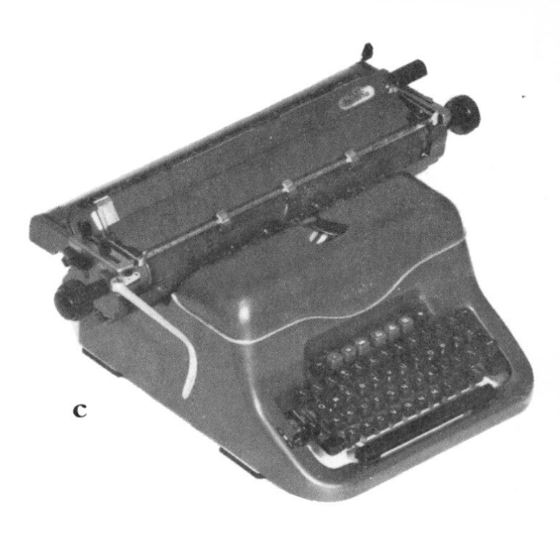

c

5.19 Triumph Standard Typewriters
- a Model 2 of 1911
- b Model of 1919
- c Matura Model of 1950
- d Matura Model of 1968

a

b

20 Triumph Portable Typewriters
- a Perfekt Model of 1928
- b Perfekt Model of 1929

ee pages 173–5

21 Adjusting a Remington Standard Typewriter made under licence during 1933 in the Zbrojovka Works in Czechoslovakia

a b

5.22 Remington Model 12 of the Z Series (left-hand carriage return) of 1929–34. Made under licence

5.23 Zeta Typewriters (also known as Consul or Diplomat in U.K.)
a Standard Model 1501 of 1948
b Portable Model 1511 of 1950

produced by a bicycle and motor cycle manufacturing company, founded in 1896 in Nuremberg. Triumph and Adler have always been closely associated but it was not until the majority of the shares were acquired by Max Grundig in 1957, that the foundation for Triumph's present strong position was established.

In 1927 Adler and Triumph standard machines were identical for a period of time.

In 1969 Triumph became integrated with Adler when the whole of the company was, together with Adler, taken over by Litton Industries Incorporated. Today they are T.A.V. (Triumph-Adler Vertrieb GmbH), a division of Litton Industries Incorporated, and almost

a b

5.24 Consul Typewriters
a Standard Model 1502 of
b Standard Model 203B of which incorporates a tw colour ribbon and fast f

5.25 Assembly of the long carriage into a Consul Model 203D Typewriter in 1971

all the machines are identical and sold in most countries in the world under both names. In America, however, only the 'Adler' machine is sold and the 'Triumph' name is not used at all.

There was a 'Triumph' typewriter of American conception which commenced production in March 1907, and ended in October 1907, but this had no connection whatsoever with the present production of T.A.V. which controls Triumph-Adler distribution throughout the world.

There is no need to illustrate the later models from 1969 or thereabouts as the machines are identical with 'Adler'.

ZETA AND OTHER TYPEWRITERS MANUFACTURED BY THE ZBROJOVKA WORKS IN CZECHOSLOVAKIA (SEE FIGURES 5.21–5.32)

The ZBROJOVKA Works at BRNO, Czechoslovakia, started life in 1918 and the company originally concentrated most of its resources on the manufacture of rifles and machine guns. The most famous of these was the ZB26, a light machine gun used by the British Army and known as the BREN gun (BRno—ENfield).

In 1946, with the arrival of peace, the company developed a wide range of products which included not only typewriters but machine tools, car accessories, engines, computers, teleprinters, etc.

Typewriter production had first started in 1932, when they assembled under licence the Standard Remington 16 (sold under the name Remington Z) and the portable Remington Streamline.

The relations between the two firms were severed during the war but were renewed immediately after, in 1945.

At the same time, Zbrojovka were spending time and money on research and development of typewriters of their own design, and their first typewriter was a standard machine called 'Zeta' 1501 which continued in production until 1958, when it was replaced by the 'Consul'

a b c d

5.26 Consul Travelling Typewriters
- a **Model 1531 of 1959 was the smallest model made; it had no tabulator, one small colour ribbon, and forty-two keys**
- b **Model 231.3 of 1969 with no tabulator, one-colour ribbon, and forty-four keys**
- c **Model 235 of 1970 with a column tabulator, two-colour ribbon, and forty-four keys**
- d **Model 236 of 1970 with no tabulator, two-colour ribbon and forty-four keys**

a

b

c

d

5.27 Consul Portable Typewriters
- a Model 221.1 of 1971 with tabulator, two-colour ribbon, forty-four keys and stroke regulator
- b Model 222.1 of 1972
- c Model 222.2 of 1972 with a decimal tabulator, two-colour ribbon, forty-four keys, stroke regulator and a long carriage
- d Model 223.1 of 1972 with no tabulator, two-colour ribbon, forty-four keys and a short carriage

5.28 View of the assembly of portable typewriters in 1971 in the Zbrojovka Works, Czechoslovakia

a

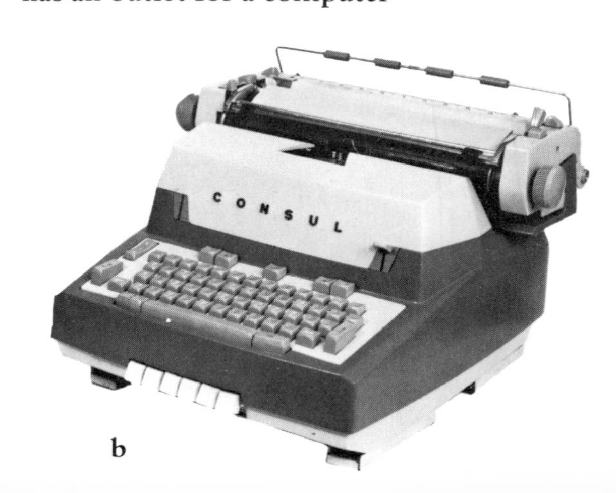

b

5.29 Consul Electrical Writing Units
- a Shows view of the assembly of Model 254 in 197 in the Zbrojovka Works, Czechoslovakia
- b Model 260 of 1971 for information entry and wh has an outlet for a computer

1502. This was modified in 1965 and was known as model 1504. Two years later the company began the production of a modern standard typewriter, the 'Consul' 203.

Their first portable came on the market in 1950, and was called 'Zeta' 1511; and other models followed in fairly quick succession, known as 'Consul' 1515, 1518, and 1519. Their present range of portables are:

'Consul' 221·1 with tabulator
'Consul' 221·2 without tabulator
'Consul' 221·1
'Consul' 222·2
'Consul' 223·1
'Consul' 223·2 } these show slight variations

In 1959 the company started manufacturing travelling typewriters. These too bore the name 'Consul'; the first model was the 'Consul' 1531, followed in 1965–7 by the 'Consul' 1532, and subsequently by models with an increased number of keys (now forty-four). These were the 1503, 1504 and 231·2. The latest models are the 'Consuls', 231·3, 235, and 236 in various colour combinations.

The company's first electric typewriter was the 'Consul' 242 and was introduced in 1964. Five years later it was replaced by the 'Consul' 243.

Consul 253 Organization Automat of 1972 for the debugging of electronics

5.31 Consul 261 Computation Automat of 1971

2 Teleprinters (made under licence from Siemens)
a Page Teleprinter T 100 of 1971
b View of an adjustment to the Page T 100 in 1971

a

b

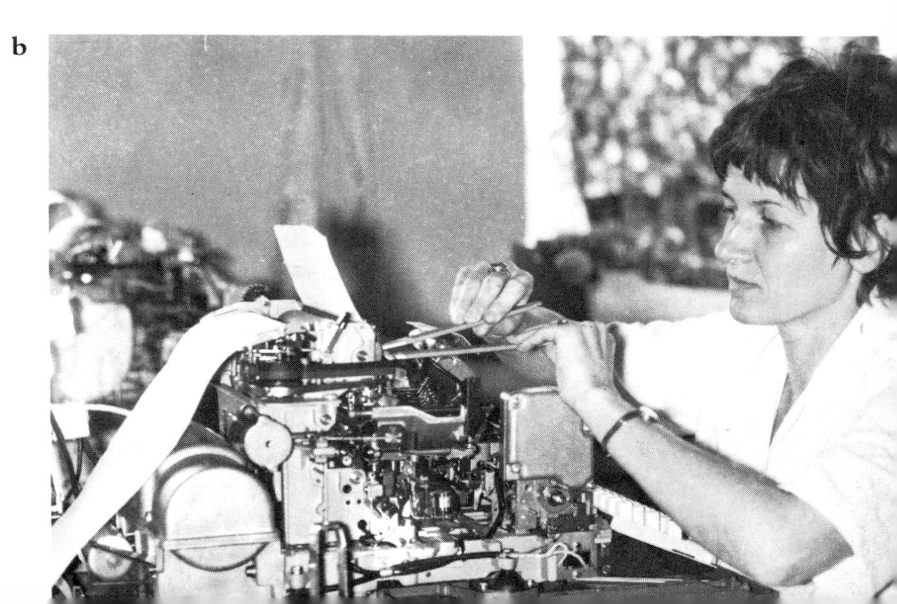

CHAPTER 6

Electric and Special-purpose Typewriters

ELECTRIC TYPEWRITERS

The inventors and manufacturers of electric typewriters have met with mixed fortunes. Some of their machines were good but others were well below standard and raised false hopes.

The object of the introduction of electricity to typing has been to secure a more even impression, greater speed, and consequently, better work and more of it. The first attempt to produce an electrically driven typewriter was probably that of Giuseppe Devincenzi who, while living in London, obtained a patent for an electric writing machine on 30 June 1855. This machine had an upright type-wheel on the circumference of which the letters were arranged. The figures were depressed by a hammer which was inside the wheel and was pressed down on the paper-carrier.

On 13 November 1871 George Arrington of Washington and Thomas A. Edison obtained a patent for an electrically driven typewriter, also using the type-wheel principle.

In 1879 C. L. Driesslein of Chicago claimed to have invented an electric writing machine. Pastor Malling Hansen equipped his 'writing ball' with an electric carriage runner. This increased the operational value of the machine considerably and it became standard practice to include it in almost every 'writing ball'.

The Magnetographe of de Neufbourg was made in 1877, and in 1892 Dr. Thaddeus Cahill was granted a patent for an electrically driven writing machine. In 1901 the Cahill Typewriter Company was founded in Washington. They marketed a machine on the style of the blind-writing Remington (understrike) and it worked with a magnet which raised the sub-levers. The spacebar and line-spacer were also electrically operated. Of this machine only forty models were made in eight years. Altogether it cost the company $157,000. The Company collapsed in 1905.

In 1899 the Hungarian, Arnold Yeremias, went to Berlin to begin the manufacture of a visible writing machine with a two-coloured ribbon, electrically operated. It was a machine of the greatest simplicity and required, it was claimed, only one-fifth of the number of parts incorporated in typewriters at that time. The depression of the key needed a touch of only 1 mm and the electrical energy was said to be produced by two small elements.

In 1900 Jules Duval, who lived in New Orleans, invented a machine which had a key for each character. Over the keyboard was the electric spring with which the letter to be written was to be touched; (it worked on the indicator principle). The letters were arranged in three rows on a metal wheel and the machine incorporated a magnet which caused the reproduction of the letter by striking the wheel. It was called the 'Express' and the letters were arranged in alphabetical order.

In 1900 news was released of the electric 'Germania' made by Sundern. Both visible and blind-writing models were made and exhibited at the Paris World Exhibition. Both machines had a magnet to attract each type lever but the carriage was not electrically operated.

In 1901 Heinrich Kochendörfer in Leipzig produced an electrically driven thrust machine. It worked with accumulators and needed only a touch of 2 to 3 mm. It is not known whether it was ever really used in practice.

Similarly, a machine by Dr. Faber of Berlin, the 'Electrograph' also used this principle. It had type levers which were arranged vertically in the manner of the type ball. The inventor reported in the same year that the experimental model he had used could produce work with an impression suitable for making a carbon copy with a view to reproducing what had been written on a hand press. This machine was never produced satisfactorily.

In 1902 the electrically driven 'Blickensderfer' was announced. This seems to have been the very first electric typewriter produced commercially and unfortunately none seem to have been preserved for posterity to admire!

The motor was screwed to the back of the machine. It was operated off what was then DC mains. The keys, touched lightly, caused rotation of the type-wheel and movement of the keys against the rotating shaft provided the electrical power. The backspacer, line-space mechanism, and margin stop were all operated by electricity. It was stated that on the left of the type-wheel there was 'an automatic apparatus' with the help of which it was possible to draw vertical, horizontal, and diagonal lines or strokes in the colour desired.

It would appear that on certain models some means of generating electricity by foot was provided so that the machine could be maintained in operation should the main electricity supply fail. Information concerning this is, however, scarce. Only a few found their way to Europe, and some are known to have come to England. Manufacture was unfortunately abandoned due to the sudden death of Blickensderfer in 1917.

The Yetman came next in 1903 and this was made in the Monarch factory in Syracuse. It was like a teleprinter, and a copy of the machine bearing the inscription 'Smith Visible' is on show in the Milwaukee Public Museum. Nothing more was heard of this machine after 1910.

In 1901, Ennis designed a machine to sell for $100. It was a type-wheel machine on which the keys struck down onto the roller from above. It was never produced commercially, but G. H. Ennis received $100,000 for it. Surely the greatest reward for any effort in typewriter history!

6.1 Blickensderfer Electric Typewriter made in U.S.A. in 1902. The machine, belonging to Group 2, was the first electric to be produced on a production line; however no models appear to exist

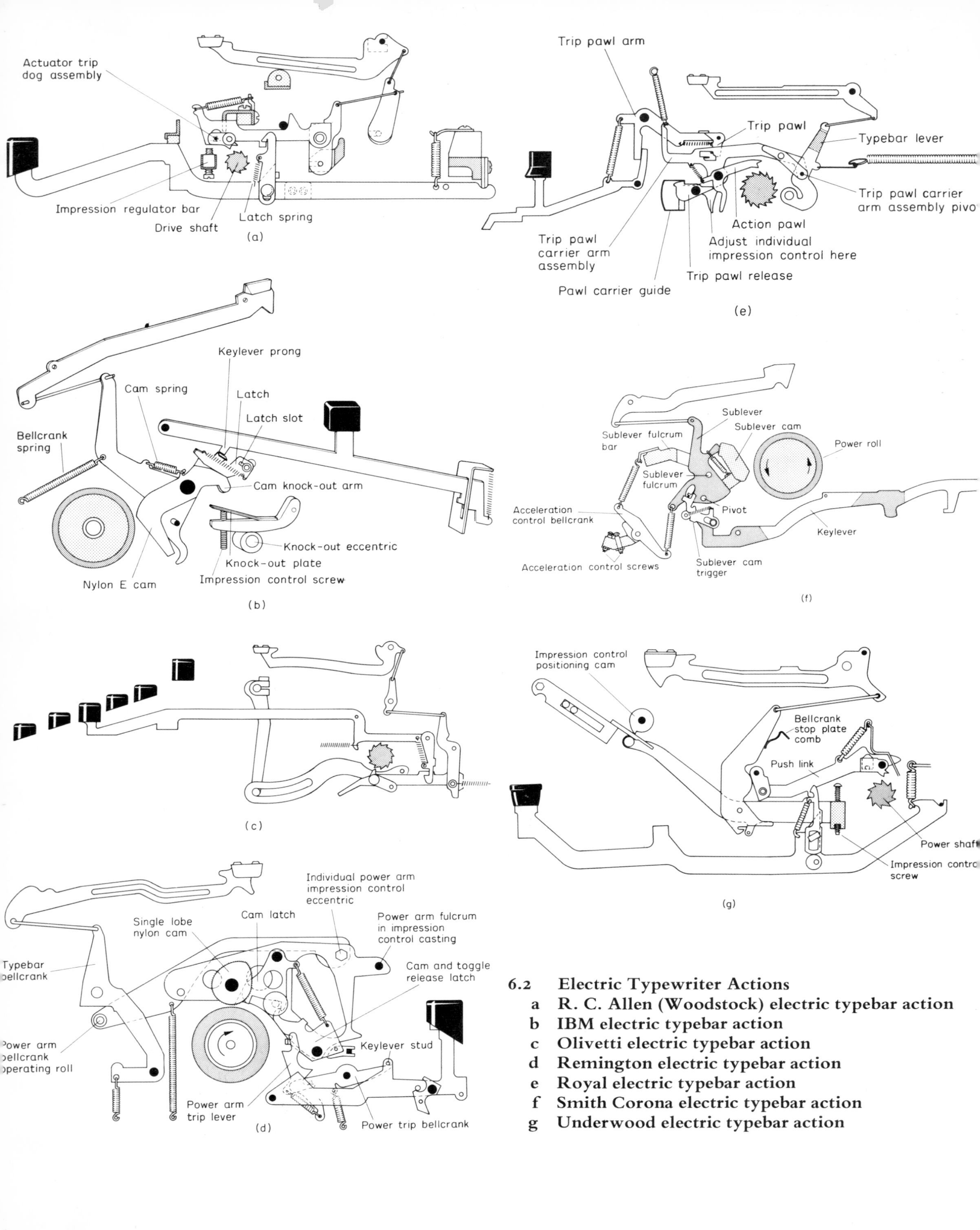

6.2 Electric Typewriter Actions

- a R. C. Allen (Woodstock) electric typebar action
- b IBM electric typebar action
- c Olivetti electric typebar action
- d Remington electric typebar action
- e Royal electric typebar action
- f Smith Corona electric typebar action
- g Underwood electric typebar action

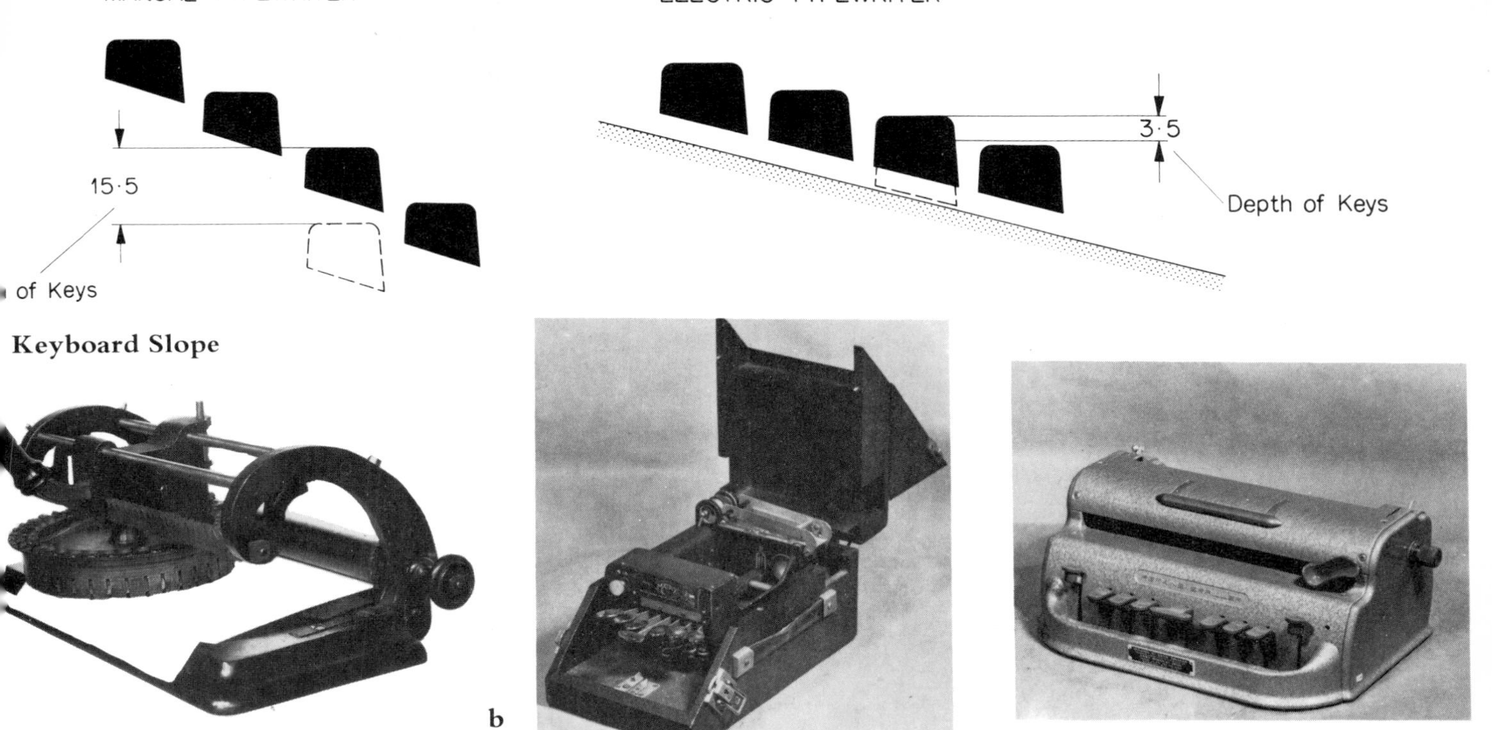

Keyboard Slope

Writing Machines for the Blind (see page 183)
Martin Typewriter of 1862
A Braille Writing Machine Matrix Model F No. 1,221 of about 1950
A modern Braille Writing Machine called the Perkins Brailler, made in U.S.A. in 1972

Noiseless Typewriters (see pages 183–5)
A three-bank machine of 1904 made by the Parker Machine Company, Buffalo, U.S.A. It belongs to Group 4 and has three rows of keys
Remington X or CX (Canadian) Model 10 of 1934–9
Remington EX (English) Model of 1965

6.6 Cryptograph AB is a Swedish Cypher Machine of 1929 (see page 185)

In 1906, the Electric Power Typewriter Company in Bradford, Canada, produced a machine for $600. In 1910, Joseph Taylor of Rochester made an electric machine. In 1912, Oscar Fischer in Berlin produced a machine similar to the Crandall-Moya which was called 'Legato' to be built by the German Munitions Factory in Berlin or by Seidel & Naumann, who manufactured the 'Ideal'. It never appeared on the market.

In 1917, efforts were made to electrify the 'Picht' machine for the blind, but attempts failed because of the high cost involved.

There were several applications for patents in the following years but the only successful one was the Mercedes 'Elektra' in 1921. The work had been carried forward to such an extent that the machine was ready in 1914, but it was delayed because of the war, and in 1919, work was resumed and it was commercially produced in 1921. The principle was copied in the Woodstock 'Electrite' and led to the construction of the 'Electromatic'.

The original Mercedes 'Elektra' had the same appearance as the standard machine except that on the right-hand side was the little electric motor. This took its power from the mains, and for every kind and voltage of current there was a special motor. The keys and shift keys were driven electrically and only a touch of 2 mm was needed to make the impression. The electric current was immediately cut off when the key was lifted and the impression produced by the electric power raised by touching the keys. The impression remained constant, regardless of the weight of the operator's touch. The carriage return was not originally electric. This followed later. A locking device prevented a second key from being depressed until the first one returned.

In 1927 an improved version was produced, and in 1933 the Model 3 with a smooth and silent touch was introduced. In 1924 the Woodstock 'Electrite' was first sold. It was constructed by O. A. Kokanson and was provided with a motor for all kinds of current. Touch

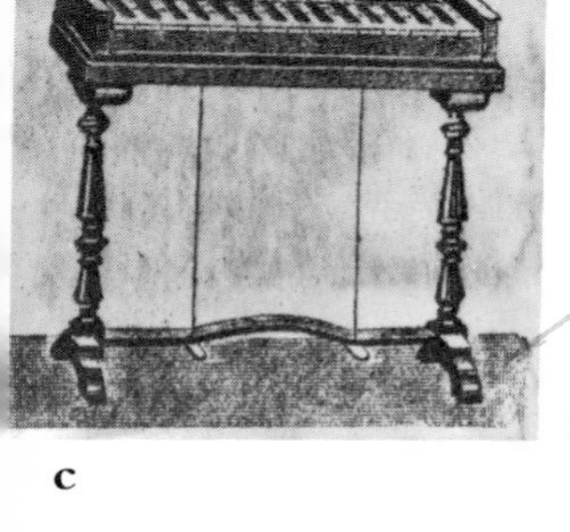

c

a

b

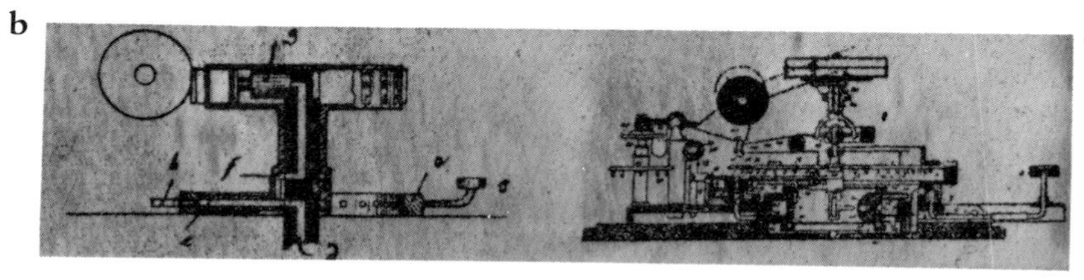

6.7 Pneumatic Typewriters which, to the best of the author's knowledge, were never produced commercially, and none exist (see page 186)

a Model of 1892 produced by Marshal A. Wier in London

b Soblick Type Wheel on Axis about 1898

c The only existing drawing of the Moser Machine of 1900

could be regulated at will, and it could also be used without electricity. The carriage return was manually operated.

In 1925, the first model of the Remington electric was shown at the Exhibition of Office Machinery in New York, and was first sold in 1927. In addition to the typebars, the shift key, backspacer, and tabulator were also electrically operated. The keyboard was flatter than on the standard machine and a light touch was sufficient to operate it. It was also possible to couple machines together, using one as a master with various 'slave machines' all repeating the action of the master simultaneously.

In 1927, the electric 'VariTyper' was produced. It was a development of the original 'Hammond' and more than 300 type styles were interchangeable for fifty-five languages. Brochures, print, etc. could be made from these and printed by offset litho. In 1930, the

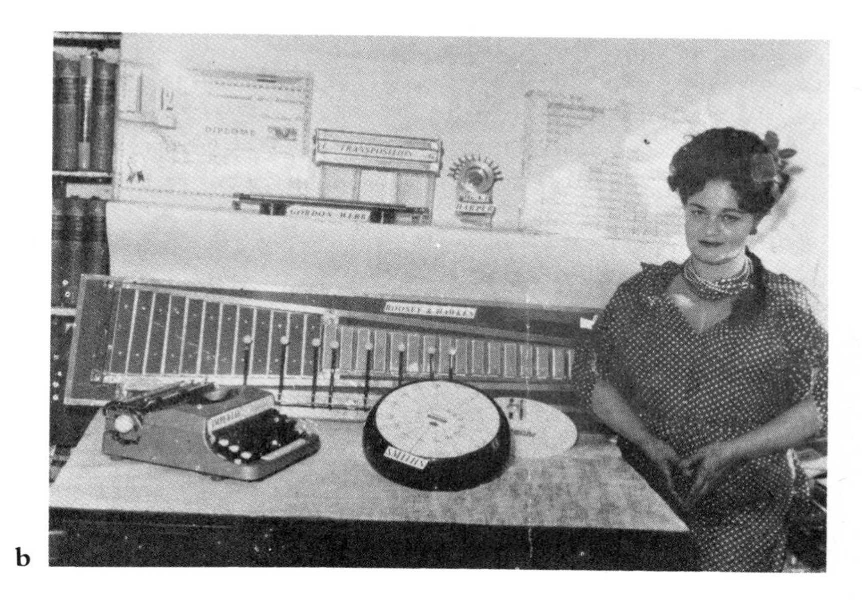

b

6.8 Music-Writing Typewriter (see pages 186–7)
a Lily Pavey, from London, ingeniously adapted this typewriter so that it can be used by a typist horizontally and vertically for music
b The inventor together with several music teaching aids sold with the Typewriter

'Electromatic' appeared, the story of which is found at the beginning of the history of the IBM.

A few years later, the Remington Company, having gained some knowledge of experiments to electrify typewriters, decided to convert 2,500 of their own machines. This was the Model 12 with right-hand carriage return and was the first step towards the production of electric typewriters in quantity.

The watchmaker George Pellaton, brought out a small electric typewriting machine in 1932, without keys. The carriage return and line-spacing were also electrically operated. The machine never appeared on the market. An attempt was made to electrify the 'Mignon'; it was never produced commercially because of the high cost involved.

Prior to World War II, electric typewriters did not achieve much public acceptance, probably due to the fact that the many early models had mechanical faults which prevented customers from taking advantage of the benefits of electric operation.

After World War II in 1946, the Underwood Corporation introduced an electric typewriter.
In 1947 Remington Rand did the same thing.
In 1950 Royal introduced its first electric typewriter.
In 1953 R. C. Allen Company introduced the Woodstock 'Electrite'.
In 1954 Smith-Corona introduced its first electric typewriter.
In 1956 Olivetti introduced the 'Lexicon' Electric.
In 1957 Adler introduced the L & S Model.
In 1957 Triumph also introduced its first electric.
In 1958 Olympia introduced its first electric typewriter.
In 1960 Imperial commenced production of an electric machine.
In 1961 IBM introduced an electric machine which departed radically from the accepted typewriter design. This was called the 'Selectric'. It was based on a mechanical principle used in the Blickensderfer writing machine of the late nineteenth century, with many

modifications, incorporating a globe-like type element instead of typebars. The 'Selectric' was unique in that the ball head moved and the carriage remained stationary.

The public were very slow at first to accept the electric typewriter. This was partly due to the unreliability of the machines and partly to the reluctance of the typist to accept electric typewriters. An odd state of affairs existed. Ladies who were quite prepared to use almost anything electrical in the home were frightened in some way that they would be electrocuted by a typewriter. It is ironic that ladies, who had their hair washed and who placed their wet heads against an electric dryer whilst soaking their hands in water, and were a fraction away from almost certain death should some minor fault occur, were fearful of the results of using an electric typewriter.

Figures show that from the end of the war in 1946 until 1952, the number of electric typewriters purchased in the United Kingdom was negligible.

In 1952 there were 450 machines purchased; most of these were IBM. By 1966 the annual total purchased was 12,700. By 1965 the figure had risen to 29,000 per annum, and by 1972 it had increased to 80,000.

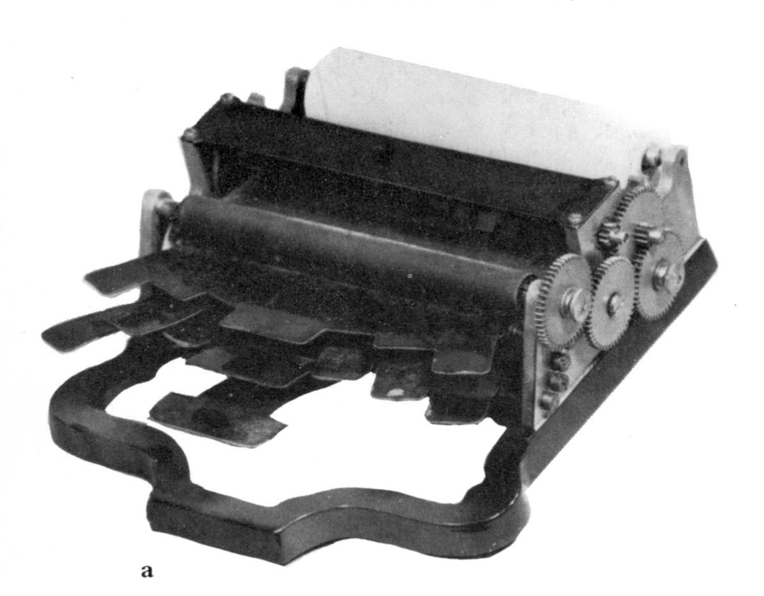

a

c

b

6.9 Shorthand Writing Machines (see page 187)

a Steno-Typer prior to 1906

b Stenotype of 1911 made by the Stenotype Company, U.S.A.

c Shortwriter of 1914 made by the Shortwriter Company, U.S.A.

As reliability and public acceptance grew, so the age of the Electric Typewriter dawned. All the leading manufacturers produced electric machines, and the number of electrics rapidly began to overtake the manual market, even on the portable side. Also a new range of machines has spread which is known as the 'Compact' Electric. These are not fully standard machines and not portable, but in between, and are very acceptable to small businesses that feel a need for something larger than a portable, yet do not have sufficient work for a fully electric standard office machine.

It is almost a foregone conclusion that with the revolutionary developments which are about to take place, the manual typewriter will eventually become a thing of the past, and all machines will be electric, working on completely revolutionary principles, and 'steam age' typewriters as we know them will disappear completely in the not too distant future.

Keyboard Slope: with the introduction of electric typewriters, typing fatigue has diminished. The keyboard slope of 15° which we have today is the result of extensive studies and tests. The extent to which the keyboard slope has been changed can be seen in Figure 6.3.

WRITING MACHINES FOR THE BLIND

Early attempts to produce a writing machine were mostly directed towards the production of machines to enable blind people to communicate with other blind people, with the aid of piano-shaped keyboards which would emboss writing or symbols onto paper to enable the receiver to feel the characters.

In 1834 Braille perfected a system of six dots which could be used in various combinations and, based on this system, numerous typewriters have been manufactured and still are manufactured to enable the blind to communicate with each other.

In addition, most typewriter manufacturers produce a typewriter suitable for blind people to use. This necessitates their learning the keyboard. The modifications are very small, requiring only four central keys to be made distinguishable from the others and a central location device for the carriage. In most cases the scale is numbered in Braille.

Three illustrations of Braille writing machines are shown in Figure 6.4.

NOISELESS TYPEWRITERS

The 'Noiseless' typewriter had its origin in the association of Mr. W. P. Kidder and Mr. C. C. Colby of Quebec, Canada, which began in 1891. They developed a close personal friendship which lasted many years.

Mr. Kidder had invented printing presses and also two typewriters, the 'Franklin' and the 'Wellington'. He turned his attention to considering problems involved in producing a typewriter which would function without noise. The mechanical difficulties involved were such as to make a heavy demand upon time and money.

Some $500,000 were invested in developing the 'Noiseless' typewriter during the years 1904–09 and he began manufacturing in the United States in 1912.

The first attempt to put this machine on the market was disastrous, and in 1912, one of the Company's engineers invented a different type action, and by 1915, the new product was being tested out carefully in small quantities.

In 1917, the Model No. 4 was produced, and this was the first 'Noiseless' typewriter in its fully commercial and satisfactory form. The three-bank 'Noiseless' portable was produced in 1921, and much of the Company's effort was devoted to producing a portable 'Noiseless' machine.

'Underwood', 'Continental', 'Ideal' and 'Olympia' had different ideas, and they produced machines in padded wooden boxes, but they had very little success.

In 1924, the manufacture of the 'Noiseless' typewriter was taken over by the Remington Company; thereafter it was known as the Remington 'Noiseless', and in 1925 the Model 6 appeared with four rows of keys and what had come to be accepted as a normal four-bank keyboard. This was also sold under the name of Smith Premier 'Noiseless'—Model 61.

It would appear that by mutual arrangement the Remington and Underwood Typewriter Companies exchanged patents, which allowed Remington to produce the 'Fanfold' machine and Underwood to produce a machine almost identical with the Remington 'Noiseless'.

In 1933, the Underwood 'Noiseless' was altered, the front construction being much lower, and the pressure indicator being placed on the right side of the machine, with various modifications. Both Underwood and Remington may have found the demand for 'Noiseless' machines declining. They made fewer and fewer.

As far as can be ascertained, Underwood ceased production of the portable version in 1939, and of the standard version about 1947. Remington, however, continued for some years and ceased production somewhere about 1968, purely due to lack of demand.

In Germany, between 1921 and 1943, the Continental 'Silenta' which consisted of 4,000 separate parts, was produced by Wanderer-Werke of Chemnitz. In spite of this complicated arrangement, it weighed only 15 kg. It was without question a first class machine, and the best 'Noiseless' machine ever produced, but the factory was heavily bombed during the war and although certain 'Silenta' machines appeared for some years after the war, it was apparently decided to discontinue this machine due to the general situation, shortage of materials, and lack of demand.

Table 15

IMPORTANT RELEASE DATES OF NOISELESS TYPEWRITERS

Model	*Release Date*	*Characteristics*
Cam-Model	1909	Double shift. Operated by cam action. Not perfected.
Model 1	1914	Pendulum weight. Double shift. Great step forward in adoption of pressure printing.
Model 3	1917	Double Shift. Improvement over previous models.
Model 5	1925	Double shift. Embodying many improvements as a result of Remington Noiseless merger.
Model 6	1926	Standard 4-row keyboard Two-letter prefix, first Q. code 5, 6, 7, 8 and all letter. In 1929—X 100,000.
Model 10	1934	Serial X 300,000 and up.
Model 11	1937	Variation of Model 10. Serial T 10,000 to T 72,000
De Luxe Grey Model	1949/1969	Improved grey model. Serial X 705,000 and up.

Production ceased in 1969 in Scottish Plant.

'Olympia' in Erfurt also made an attempt to produce the Model 8 with reduced noise. They split the platen into noise reducing chambers and the type levers were normal but surrounded by felt padding. Attempts were also made to reduce the noise of the shift and carriage mechanisms. A table was then produced with a hinged glass cover which covered the machine completely. Other German and American manufacturers have also experimented with various kinds of pneumatic actions but without any real success.

In retrospect, it is difficult to understand why so much time and energy were expended upon creating a 'Noiseless' typewriter. Manual, and certainly electric typewriters are as noisy now as they ever were.

For some obscure reason the operators of typewriters are more concerned with 'touch' and general operational efficiency than noise reduction. The simple hard unpalatable fact which it has cost manufacturers hundreds of thousands, if not millions of pounds, to find out, is that for some unaccountable reason, a 'Noiseless' typewriter is something that people just do not require. Operators like to hear the sharp crack as they type. In this way they feel they are achieving something. There are, of course, the few who feel the need for a 'Noiseless' machine, but they are certainly in the minority, as the sales of typewriters clearly indicate. 'Noiseless' typewriters are no longer produced anywhere in the world.

CIPHER MACHINES AND SECRET WRITING MACHINES

Herodotus was one of the first to use secret writing, and Plutarch gives details about the secret writing used by the Spartans. Julius Caesar had an alphabet of his own.

Cryptography has been used by mathematicians, abbots, statesmen (e.g. Cardinal Richelieu) and various authors (e.g. Edgar Allan Poe and Jules Verne); also in later days Field Marshals devised various ciphers for security reasons.

In World War I, the use of codes and ciphers was extended and Army Headquarters employed decipherers to unravel enemy systems. The first pioneers of secret writing machines seem to have been Dujardin in Lille and Pape in Paris.

Movable types were inserted in a different order on the keyboard and then written by the machine which produced an unintelligible text. This could then be deciphered when it was copied by the receiver with a machine which had been altered in the same way.

In 1906, Hubert Burg of Lorraine, made a machine which at the same time produced a copy of the normal text, and a copy of the secret text. One of the type-cylinders was provided with a shift mechanism.

In 1919, the International Cipherwriting Company of Chicago produced a cipher machine in conjunction with the 'Hammond' which was said to be designed by a Frederick Sedgwick. The key to the cipher was provided by a system of needles in the small holes of a round plate. Many models were designed and used secretly by various governments.

In 1927 'Cryptograph' Accounting and Tabulating Company of Great Britain Limited, London, designed a cipher machine.

In 1929 a Swedish firm also produced a 'Cryptograph' designed by Engineer Ferno Teknik of Stockholm. This cipher machine worked with an electrical element in such a way that by pressing down a key a lamp was illuminated.

At the beginning of World War II, the Japanese believed they had an unbreakable cipher. This was clearly demonstrated in the film *Tora! Tora! Tora!*, which showed how two IBM machines were so linked together that the same letter never appeared in relation to another letter unless the receiver had the mechanical insert to control the two electrically linked machines.

However, the Americans succeeded in breaking the Japanese cipher which was a great achievement and throughout the war they were able to read all messages. This was a great contribution to the war effort especially as the Japanese believed that everything they sent was secret!

Typewriters increased the possibilities of code writing and the linking of electric typewriters was an even greater step forward. But with time and determination all codes and ciphers have always been broken.

Very little is known about present developments which of course are highly secret, but it is generally believed that communications of this nature are now typed in a cipher or code form. The message is recorded at pre-arranged speeds on tape recorders, speeded up and re-recorded *ad infinitum*, so that texts amounting to thousands of words are transmitted in a matter of micro-seconds three or four times, and then played back on various tape-recorders at pre-arranged speeds.

PNEUMATIC TYPEWRITERS

Various attempts were made from 1891 onwards when Marshal A. Wier in London, produced a typewriter with a pneumatic action. The object of such a machine was to eliminate the hard work involved in typing, and to reduce the noise and increase the speed. It was also thought to be a substitute for such power as electricity.

One of the disadvantages of pneumatic machines has always been typebars that did not return fast enough, and although this problem could most likely have been overcome the fact is, it just seemed to present insurmountable difficulties.

It would appear that the last real attempt to manufacture a pneumatic machine was made in 1914, by a man called Juan Gualberto Holguin in Mexico. This machine was known as the 'Burbra', and used compressed air cylinders as a source of power. In spite of much time and money spent on the production of compressed air typewriters, very little result of any importance has ever been achieved.

There are reports of various designs of pneumatic typewriters having been produced by large organizations, both in America and in Germany in recent years. Most of these consisted of an electrically propelled plunger which compressed oil in a tube, fired the typebar forward in a sharp thrust, had the advantage of being very quiet and also eliminating most of the moving parts of the conventional machine. The idea seems to have been abandoned, due to the high cost or probably to lack of interest.

As explained elsewhere, the quiet action, for some obscure reason, is not popular with typists. Theoretically, they should be pleased to operate in complete silence. In practice, typists like to hear the noise of their machines and seem happier with either manual or electric typewriters, regardless of the noise they make. As the customer is always right most of these costly ideas have been abandoned. Usually when typists complain of noisy typewriters it is because they dislike the machine for other reasons. There could be some future development in pneumatic machines, but it is highly improbable.

MUSIC-WRITING TYPEWRITERS

As early as 1745, experiments were being carried out with music-writing machines. No doubt the production of typewriters for the blind with its emphasis on the piano keyboard gave rise to machines which would print music. There are a number of references in early works to machines dating from 1771 to 1808. Progin produced a machine in 1833. Many experimental models were produced. Also a considerable amount of time and money was expended to modify existing typewriters, without much success.

Prior to World War II, a number of serious attempts were made by both 'Remington' in 1937, and 'Continental' in 1939, but for some reason these machines did not attract any great attention.

As late as 1950, the Imperial Typewriter Company produced the Pavey 'Musigraph' machine which was abandoned due to cost and lack of general interest. It was a modification of the Model '70'. The market for such machines has always been very limited, and the photocopying machine made the copying of music so simple that complicated typewriters were completely unnecessary; moreover, anyone composing an original score, would much prefer to use a piano and a pen.

SHORTHAND WRITING MACHINES

An incredible number of shorthand writing machines have been produced, the first reputedly having been made in 1827, by the French librarian Gonod of Clermont-Ferrand. Almost every few years from then onwards someone made another attempt, mainly concentrating upon, once again, the piano keyboard, and later with individual keyboards. Usually these machines were sold in conjunction with some special system, and rented, but in most cases the shorthand typist has preferred to use her own particular brand of shorthand. The shorthand machine has a very limited application today as its place has largely been taken by dictating machines.

TOY TYPEWRITERS

Several attempts have been made to produce toy typewriters for students and children. These have had very little success and a limited sale.

The first recorded machine was the Pocket Typewriter produced in 1887. This was followed by the 'Simplex' toy typewriter in 1892 which was constructed of solid brass and was extremely well made. It was produced in America and the manufacturers attempted to retail it for one dollar. Needless to say, the company failed and production was soon discontinued. The one illustrated (*see* Figure 6.10*b*) is in the British Typewriter Museum in Bournemouth.

The 'Lilliput' toy typewriter was manufactured in 1955 in Croydon, England, and the 'Petite' toy typewriter was made in 1960, and manufactured in Nottingham, England.

Other toy typewriters are manufactured in Japan.

All these machines are sold in relatively small quantities due to the low price of real portable machines.

a

6.10 Toy Typewriters

a The Pocket Typewriter of 1887 is a primitive machine weighing only about 4 oz and designed by Dobson and Wyn. It was conceived as a child's toy and the characters are printed by turning a handle; the types themselves are underneath a cogwheel

contd. overleaf

c

d

b

6.10 Toy Typewriters (contd.)
b Simplex Model
c Lilliput Model of 1955
d Petite Model of 1960

REBUILT TYPEWRITERS

A few years after the production of the first machines, a market developed in rebuilt and reconditioned typewriters which still exists today.

Rebuilt machines are produced either in the factory where they were originally made, or by one of the many rebuilders who buy any make of machine regardless of its age or condition. Each machine is broken down into its component parts by specialists. The rubber parts, damaged types or type levers are renewed; in fact all worn-out parts are replaced by new ones and adjusted with precision so that the machine again writes perfectly without any 'lost motion' in the keys. The keys and the black enamel parts are newly enamelled, and the model number and other markings are sometimes changed. Such rebuilt machines operate almost as well as brand new ones, and are appreciably cheaper.

Before World War II, the demand for used machines led to a large and thriving business both in America and England and, indeed, all over the world.

Owing to the general depression in the early 1930s, many privately-owned typewriter firms were dealing almost exclusively in used and reconditioned typewriters, for new ones were almost unsaleable. In fact, at one time, the problem became so acute that when manufacturers acquired second-hand machines, they would break them up to prevent them from being returned to an already over saturated market.

In contrast, other manufacturers such as Royal and Imperial set up factories to rebuild their own machines as quality reconditioned models, at a lower price, and thus maintained their good name. In various parts of the world used machines were reconditioned on factory lines to supply the typewriter trade.

Between World Wars I and II, the design and manufacture of machines varied little from year to year. Manufacturers introduced small alterations but the basic machine remained the same.

After World War II, most manufacturers continued with their pre-war machines until they were able to design something new required by the expanding market. This, however, mainly took the form of changes in the outer casing, and the introduction of colour; the basic machine varied very little indeed.

The first electric typewriters produced after World War II were only manual machines electrified, and this brought all kinds of problems. It was only in the late 1950s and early 1960s that manufacturers started to design new types of machines which were essentially electric typewriters, instead of manual machines with a motor attached. Once this fundamental change was achieved, most of the problems surrounding electric machines disappeared, and we now have modern manufacturing processes and new designs of portable, electric, and manual machines. Though new today, they will presumably soon be sold as reconditioned and rebuilt machines.

In view of the possibility of post-war over-production, it was agreed that all 'lend-lease' typewriters supplied by the United States to the allies were to be destroyed. Consequently, after the war the ludicrous situation arose all over Europe, of a desperate typewriter shortage, whilst at the same time some brand-new and reconditioned machines were either thrown into the sea, smashed by bulldozers, dropped down disused pit-shafts, or otherwise destroyed.

The heavy bombing in World War II not only destroyed many typewriters in towns, but also a large portion of the European factories in which they were made, and as the typewriter production in most countries had virtually ceased for four or five years, a world-wide shortage arose and between 1946 and 1952, almost any machine was saleable. All kinds of wrecks that had survived were dragged out of cellars and attics and sold at fabulous prices. It took approxi-

a

6.11 Chinese Typewriters

a Lin Yutang watching his daughter type on the machine he invented in about 1950 called the Minghwai

contd. overleaf

mately six to seven years to recover from the war, but once factories were again in full production, the over-production of typewriters soon became apparent. In spite of this, however, there is still a thriving business in rebuilt machines.

CHINESE TYPEWRITERS

In addition to the Chinese typewriters being manufactured in Japan, two attempts have been made to produce other Chinese typewriters. The first was by an Engineer, Chou Hou-K'un who marketed a Chinese typewriter in 1911. About 1950, Lin Yutang invented a typewriter which, by superimposing one character on another, could produce sufficient Chinese characters somewhere in the region of 5,000. It is called the 'Minghwai' (Figure 6.11*a*).

The photographs shown in Figures 6.11*b* and *c* are of a Chinese typewriter manufactured in Shanghai.

Very little is now known about the production and manufacture of this machine but many thousands of characters are used and the general principle appears to be that of the Japanese typewriters of a similar basic type.

b

c

6.11 Chinese Typewriters (contd.)

b We believe this machine is made in Shanghai and is still in production.
There are three type cases available each one with approximately 3,000 characters

c Close-up view of some of the 3,000 characters mentioned in Figure 6.11b

CHAPTER 7

Machines Not Lost—but Gone Before

At the turn of the century there were thirty manufacturers producing typewriters in America alone, to say nothing of the many others in Europe. Looking back, it is very difficult to see why the shape and basic design eluded so many of them for so long.

The real story of all typewriters can be summed up very simply.

Firstly, there is a dedicated man with an idea, followed by a second man, usually someone with his eye on the main chance, who sees a vast fortune to be made out of an inventor's brains. Together they reach the poverty line in their desperate attempts to produce a writing machine. At this stage they enlist the support of a third person—usually someone with considerable funds—who is engaged in making something else.

In time this third person becomes impoverished and defeated by the utter impossibility of producing a writing machine in a workable form at the right price. After all, even now, after a hundred years of typewriter production, widely experienced and highly-capable manufacturing firms with a reputation for making excellent machines suddenly produce a troublesome model.

Finally, if they manage to get this far—and how many have perished along the road it is not known for sure—the fourth man enters the picture. He is usually someone of considerable means, successfully and profitably manufacturing bicycles, sewing-machines, rifles, cameras, motor cycles or motor cars, and he decides to take the plunge to disaster. By the time he has invested considerable sums of money, typewriters begin to roll off the production lines with ever increasing speed. Those who are left of the gallant band have now reached what they believe to be the top of the final hill and gaze at the vast world market that awaits their production.

But alas, to their horror and dismay, the world is not waiting for the writing machine, and the problems and torments of design and manufacture fade into insignificance compared with the struggle that lies ahead to convince people they should throw away their pens and use a typewriter. Their difficulties in selling typewriters can be greater than all their other problems combined, and can seem well nigh insuperable when in competition with other typewriter manufacturers similarly engaged in a frantic search for customers. Such is the sad tale of typewriters that have come and gone!

Most of the people who make and sell typewriters today would probably much rather be in some other business making something else, for the competition and pressures are enormous.

7.1 View of the office of the General Electric Company in 1882

It is very doubtful whether anyone has ever really made any substantial sum of money manufacturing typewriters alone and managed to keep it. Those who have, have been the visionaries who, looking beyond the typewriter as an instrument solely for correspondence, saw it as a vehicle for something else, namely the basis of sophisticated book-keeping machines or computers.

Some manufacturers developed models which lasted for many years and have now ceased to exist. Their fate forms the subject of this section.

If anyone reading this is thinking about manufacturing a revolutionary typewriter, they should think long and hard for it would appear to be the quickest way of losing a fortune!

'WOODSTOCK' (R. C. ALLEN) TYPEWRITERS: U.S.A.

In 1907 Mr. R. W. Uhlrig who had previously produced the 'Commercial Visible' in 1898, first manufactured the 'Emerson' typewriter.

In 1908 the offices were removed to Chicago, and in 1910 the business was purchased by Mr. Roebuck of Sears, Roebuck, and Company, the famous Mail Order House. Later, the factory was removed to Woodstock, Illinois.

The limited speed of the machine, however, and certain weaknesses in attaching the type to the bars caused the 'Emerson' to be withdrawn from the market.

By September 1914 the No. 3 Woodstock took its place, its name derived from the town of Woodstock, Illinois, where it was manufactured. The machine was a complete success from the beginning, and it was never found necessary to withdraw any models. It was a conventional two-coloured front-strike machine with four rows of keys, segment, and bars.

The 'Annell' was introduced in 1914 and sold as a special mail-order line, but this campaign was unsuccessful and the machine was withdrawn.

In 1915 the Woodstock No. 4 was produced. By this time all the executive offices and the factory were housed in Woodstock. The machine was looked on as one of the well-known standard American typewriters, and from 1917 onwards sales were extended to every corner of the world. It is doubtful if any typewriter invaded world markets so quickly or so successfully. The demand for Woodstock machines was so huge that for some time the company had great difficulty in delivering on time.

a

b

c

e

f

7.2 Woodstock Typewriters

- a Emerson Model was an early machine produced in 1907 and was really the Woodstock
- b Annell Model of 1914 was really the Woodstock Model 4. It was sold in a mail-order campaign which was unsuccessful
- c Model 4 of 1914 had a single shift (Serial No. B 26,433)
- d Model 4 of 1915
- e Model 5 of 1922
- f Electrite Model of 1925 was one of the first practical electric typewriters. It was similar in design to the Mercedes Electrite
- g Model of 1937 (Serial No. 486,155)

b

7.3 R. C. Allen Typewriters

- a Manual Standard office machine which was discontinued in 1967
- b Electrite Model E811C of 1966–7

Table 16

PRODUCTION OF WOODSTOCK (LATER R. C. ALLEN) TYPEWRITERS

Woodstock

Prefix '8K' indicates 18 in. carriage.
Prefix '8L' indicates 22 in. carriage.
Prefix '826' indicates 26 in. carriage.
Closed-in sides commenced at 324,000.
Key-set Tabulator (as standard equipment) commenced at 437,000.
Typebar Hoods commenced at 450,000.
Basket Shift at 839,000.
Prefix 'N' used for all size carriages after 530,000.

Nos. 3 and 4	1914—1917

No. 5

Prefix B, D, and DW	1917—1921
Prefix F, H, and HN	1921—1925
Prefix N to 200,000	1925—1927
200,001 to 300,000	1928—1931
300,001 to 400,000	1931—1934
400,001 to 440,000	1934—1935
440,001 to 475,000	1936
475,001 to 505,000	1937
505,001 to 530,000	1938
530,001 to 554,000	1939
554,001 to 579,799	1940
579,800 to 601,999	1941
602,000 to 622,999	1942
623,000 to 642,999	1943
643,000 to 662,999	1944
663,000 to 729,999	1945
730,000 to 782,999	1946
783,000 to 855,999	1947
856,000 to 891,999	1948
892,000 to 999,999	1949
1,000,000 upwards	1950

Prefix '812' indicates 12 in. carriage.
Prefix '8J' indicates 14 in. carriage.
Prefix '8J' indicates 16 in. carriage.

R. C. Allen

1,100,000	1950
1,116,000	1951
1,142,000 to 1,165,999	1952

Model 600

1,166,000	1953
1,176,000 upwards	1954

Model 700

1,750,000	1953
1,768,000	1954
1,820,000	1955
1,870,000 to 1,919,999	1956

Model 800

1,920,000	1957
1,980,000 to 2,029,999	1958

Model 'A' Visiomatic

2,030,000	1959
2,100,000	1960
2,200,000	1962
2,300,300	1963
2,330,800	1964
2,350,400 to 2,358,399	1965

Model 'B' Visiomatic

2,358,400	1966
2,367,200 up	1967

Many gold medals were awarded at a contest held in Paris in 1922. The championship was won on a Woodstock at a speed of 120 words per minute. It was said in 1923 that the development of the Woodstock, as a machine of growing importance in the industry, was assured.

In 1925 the 'Woodstock' produced the first practical electric typewriter, the principle of which is the basis of most other machines in use today. It had a motor adjustment, electrically controlled stroke, and an electric carriage return. Relatively few of these were sold as the machine was ahead of its time, and the world was not ready for electric machines in quantity.

Their typewriters were produced in large quantities for the American Government throughout World War II continuously, whereas some typewriter factories ceased production and produced armaments.

In 1950 for some reason, the name of the machine was changed to the 'R. C. Allen' and production of typewriters continued until 1967.

The R. C. Allen Business Machines Inc. still exists, but they no longer manufacture typewriters; thus a typewriter which had commenced production in 1907, ended in 1967, after sixty years of continuous production. Some millions of machines were manufactured and used satisfactorily throughout the world.

THE BARLOCK: NOTTINGHAM, ENGLAND

In 1914 the Barlock Typewriter Company Limited of London purchased from the Columbia Typewriter Manufacturing Company of New York all patents, names, tools, and machinery

7.4 This is an early Barlock Typewriter which was originally called the Columbia Barlock, was produced in 1888 by Charles Spiro, and was manufactured in New York. There also appear to have been other models, and this machine was sold on various markets as the Royal Barlock, the factory having previously produced the small Columbia machine. It was a double keyboard machine of Group 6, down-strike from the front. After production ceased many of the tools were sold to Barlock in Nottingham who commenced production of the Barlock Model 16 (*see* Figure 7.5a)

a

b

c

d

e

f

g

7.5 Barlock Typewriters

- a Model 16 of 1921 belongs to Group 10, is four-bank, front-strike, and has an upward segment shift. It was the first machine produced by the Barlock Typewriter Company, Nottingham, England
- b Standard Model 19 of 1935 (Serial No. 361,904)
- c Standard Model Super 19 of 1939 (Serial No. 384,589)
- d Standard Model 18 of 1930 (Serial No. 351,506)
- e Standard Model 20 of 1943 (Serial No. 527,695)
- f Standard Model 20W of 1945 with war finish (Serial No. 616,569)
- g Model 21 of 1946 (Serial No. 702,409)

a

b

7.6 Byron Typewriters

- a Standard Model of 195
- b Standard Model of 195 (Serial No. 9,217)

for the manufacture of the Barlock typewriter. Previously there had been some assembling of machines from parts imported from America, but in 1918, the plant in Nottingham produced the machine in its entirety, and it became a British product. The new Model 16 was a four-bank Group 10 (*see* Table 17) machine with a segment shift which moved upwards, was of unit construction, and in its day, a very advanced machine. Later, it had a segment which moved downwards.

Through the years the Barlock Typewriter Company continued in Nottingham, supported by Sir John Jardine, whose main concern was the manufacturing of lace-making machinery. They produced not only the Barlock standard machine, but the Bar-let three-bank portable and the Barlock four-bank portable.

During the period 1930 to 1940, like many other factories, they had great difficulties, owing to the intense competition offered by American and German machines which had greater appeal. However, between 1933–5 good orders were received from the British Government, and when World War II started the model '20W' was produced at the request of the government who took the entire output.

After the war the Barlock Typewriter Company continued with the Model '20', followed by the Model '22', but found they were unable to continue to produce the portable machines competitively. For a few years immediately after World War II, any typewriter which could be produced was immediately sold without any difficulty at a good price, but with the renewal of competition, great difficulties were experienced.

In 1955 production of the Barlock typewriter ceased. The name was changed to Byron Business Machines Limited, and there was a change of directorship. Approximately one million and a quarter pounds were invested in an attempt to produce a first-class standard machine and the 'Japy' Frères portable was imported from France and sold as the 'Byron' portable. Later a totally new standard machine was designed and approximately 1,000 were produced.

This machine met with some little success, but as is almost inevitable with a new design, it had a number of faults. It was estimated that a further fifty thousand pounds were required to make modifications. This amount the directors were not prepared to spend. In 1958 the total assets of the Barlock Typewriter Company were sold to the Oliver Typewriter Company and Barlock ceased to trade.

Here was a typewriter of which some millions had been sold. It lasted for forty years and then vanished from the scene.

b

7.7a Barlock four-bank portable of 1939
b Bar-let Model 2 three-bank portable of 1935

Table 17

PRODUCTION OF BAR-LET, BARLOCK, AND BYRON TYPEWRITERS

Bar-Let Portable

To 10,000	1930—1933
10,001	1933—1935
20,001	1935—1936
30,001	1936—1937
50,001	1938
54,801 to 58,800	1939

Barlock 4-Bank Portable

To 2,810	1938
2,811 to 5,810	1939

Byron

Model 53

B5000	1952
B6450	1953
B8546	1954
B9534 to B9791	up to 21/11/55

Model 54

1001	13/8/53 to 31/12/53
1295	1954
3274	1955
5048	1956
6069 upwards	1957

Byron Portable

185,617 to 193,350	1954
196,954 to 210,975	1955
214,401 to 230,550	1956

Barlock

No. 16

300,000 to 326,000	1921—1924

No. 17

326,001 to 335,000	1925—1929

No. 18

340,001 to 352,000	1929—1930

No. 19

360,000	1933—1935
365,001	1935—1936
370,001 to 380,000	1936—1937

Super Model 19

380,001	1938
385,001 and over	1939

Model 20 (First produced February 1939)

500,001	1939
504,304	1940
511,224	1941
519,747	1942
527,199 onwards	1943

Model 20w

600,000	1943
604,696	1944
610,562	1945
618,150 onwards	1946

Model 21

700,000	1946
705,732	1947
712,850	1948
720,701 onwards	1949

Model 22

800,000	1949
806,879	1950
814,965	1951
823,018	1952
824,822 onwards	1953

THE BLICKENSDERFER TYPEWRITER: U.S.A.

7.8 George C. Blickensderfer (1851–1917)

George C. Blickensderfer invented and produced a small portable typewriter bearing his name. He was without question a brilliant man, well ahead of his time. He commenced manufacture in Stamford, Connecticut, U.S.A. in 1889, and in 1893 produced the first real lightweight typewriter to be sold in any quantity throughout the world. Like the Hammond, his first machines had the 'Ideal' keyboard, but in answer to popular demand he changed over to what we now know as the 'QWERTY' arrangement.

The 'Blick', as it became known, printed from a small ink roll, was a visible writing machine of Group 2 type-wheel design, and had an added advantage: a hundred different keyboards were available and as many different type-wheels, which were instantly interchangeable. It was sold in some countries as the 'Dactyle'.

An improved model appeared in 1897. It was the first typewriter to be placed in a carrier bag or box, and was thus the first portable. The machine was then constructed with an aluminium frame to save weight. The Model 8 appeared in 1907.

The Blickensderfer electric typewriter was first produced in about 1902. This is mentioned in Mare's book, printed in 1909, which also states that the New York Central Railway had

b

c

e

f

contd. overleaf

7.9 Blickensderfer Typewriters made in U.S.A.

a Portable Model of 1893 belonging to Group 2

b Model 6 of 1896 belonging to Group 2

c Featherweight Model of 1896 belonging to Group 2

d Model 7 of 1897 belonging to Group 2

e Model 8 of 1907 belonging to Group 2 and having a long platen

f Model 9 of 1917 belonging to Group 2. Production ceased in 1918, and in 1928 Remington produced and sold this machine by the name of Rem-Blick, but it was unsuccessful

g Blick 90 Model of 1919 was a three-bank machine belonging to Group 10

h

i

7.9h Blick-Bar of 1913 belonging to Group 9. In 1917 manufacture was taken over by Harry A. Smith

7.9i Blick-Bar of 1916 belonging to Group 9. It was produced by Harry A. Smith and very few were made

such a machine working with a 36 in. carriage which was interchangeable. It operated on 110 volts DC current and some models had a treadle with which to generate electricity should the mains fail. It was far ahead of its time as both horizontal and vertical lines could be drawn in two colours. Some of these were exported to European countries, including England. No models appear to exist today, but many thousands of the manual machines are still in good working order.

On 14 August 1917, George Blickensderfer died suddenly. The world had lost a remarkable man, whose two main achievements were to put the first portable typewriter and the first electric on a production line. He apparently had very little interest in anything else but typewriters.

After Blickensderfer's death the factory at Stamford became very profitable, manufacturing machine-gun parts and a gun carriage which he had invented. Later a 'Blick-Bar' machine was produced and much money was lost due to its unprofitable construction. It was purchased by a Mr. Harry A. Smith, and sold under that name. It had no connection with the L. C. Smith or Smith-Corona Company, although there was a great deal of similarity in the design of the machine.

The Company that had lasted thirty-two years encountered financial difficulties. A Receiver was appointed in 1921 and the remaining assets were disposed of.

THE BURROUGHS TYPEWRITER: U.S.A.

Burroughs is one of the best known names in the book-keeping, accounting machine, and cash register field today. They have a world-wide reputation for high quality and sophisticated products.

It is not generally known, but in 1931 Burroughs produced a manual typewriter, and in

b

a

7.10 Burroughs Typewriters
a Manual Model of 1931
b Electric Model of 1932 (Serial No. 64A 81,094)

1932 an electric typewriter. These were soon discontinued, however, as the company preferred to concentrate on the more profitable side of the business. Nevertheless, they did make typewriters for a short time; they continue to manufacture and sell successfully more and more sophisticated machinery with ever increasing success.

THE CONTINENTAL : GERMANY

In 1904 the Wanderer-Werke at Siegmar Schonau near Chemnitz which had been making milling machines, bicycles, and motor bicycles, produced their first typewriter; it was a standard four-bank machine with forward-strike (Group 10) of excellent quality and had all the features of a modern machine with four rows of keys, a tabulator, and two-colour ribbon. By 1914 the business had expanded to such a degree that the annual output was 60,000 typewriters, and then during the war, like other typewriter factories, they had to turn their attention and energies to producing munitions of war. In the year after the war they resumed typewriter production and in 1929, the 'Continental' portable was made, with another model being produced in 1933. Also in 1933 the Model '34' appeared which then became known as the 'Wanderer 35' and still later, the 'Wanderer 50'. This was the simplified 'Continental' portable. In 1938 came the fourth model known as the Model '100'.

They produced countless numbers of typewriters commercially from 1904 to 1939 which were sold all over the world. The production unfortunately came to a sudden end simply because the factory was somehow 'lost to the world'. The world is a poorer place for not having the 'Continental' typewriter in the form that it was.

The Continental 'Silenta' was really the only successful noiseless machine in that it worked on Kidder's principle, but each type was on its own lever whereas Kidder's machine used a double action. It was extremely well made and advanced in design, and although many tariffs were raised against it, the factory was successful in selling large quantities abroad, and it was one of the best machines ever produced; the same could be said of their standard and portable machines.

During World War II, the portable plant was moved to Liège in Belgium, and after the war, portables bearing the name 'Continental' arrived on the British market. They were however made with very inferior materials but as there was a great shortage of typewriters, they were sold in large numbers. Those who bought them were quickly disillusioned and disappointed, for although they were ostensibly the same machines, their poor quality was soon discovered. Consequently those who had taken over the production disappeared rapidly from the scene and there are no typewriters made in Belgium today.

The Continental Standard machine plant was one of the most advanced in the world and nobody seems to know what really happened to it. There are stories of the plant being dismantled by Russian technicians and of an entire train being lost somewhere between Chemnitz

a

b

c

7.11 Continental Typewriters

a Standard Model of 1927 (Serial No. 286,851)

b Model of 1927 which was a four-bank machine belonging to Group 10

c Silenta Model of 1934 which was a four-bank machine belonging to Group 4

Table 18

PRODUCTION OF CONTINENTAL TYPEWRITERS

Continental Standard	
280,001	1927—1929
400,001	1930—1931
460,001	1931—1933
500,001	1934—1936
550,001	1936—1937
600,001	1938
750,001	1939

Continental Silenta	
To 708,000	1934—1935
708,001	1936
715,001	1937
725,001	1938
734,001	1939

Continental Portable	
To 10,000	1930—1931
10,001	1931—1932
20,001	1932—1934
60,001	1934—1935
90,001	1936
200,001	1937
233,001	1938
255,001	1939
500,001	1946
501,368 to 508,240	1947

Continental

Tabulator Model commenced at 49,000, Nov. 1933. (Serial number (below 500,000) is on carriage ball race rail)

and Russia. It was also said that many German technicians were taken along with the plant and told they would return when the machine was being satisfactorily manufactured in Russia but there is no way of proving these stories.

Soon after the war some Standard and Silenta machines were still available in Europe. It was generally believed that these had been assembled from parts that had been stored.

However, within a year the machines were unobtainable and to the best of our knowledge there is no factory producing 'Continental' typewriters in Chemnitz or anywhere else in the world.

EVEREST: ITALY

This typewriter started life in Milan in 1922 as the 'Juventa', which was a small portable machine with a platen only 18·5 cm wide.

a

b

7.12 Everest Typewriters
a Sabb (Juventa) Model of approximately 1922 onwards. It became an Everest in 1931
b Standard Model S of 1944 (Serial No. 92,733)

When the firm that had been making the Juventa changed its name to S. A. Brevetta Bassa, it changed the name of the machine to 'Sabb', and it was from an improved model of the Sabb that the 'Everest' was developed in 1931. It was manufactured first by A. & G. Garisch and Company in Milan, and later by the firm Serio, also of Milan, although the factory where it was made was situated in Crema.

Between 1937 and 1962 Everest standard and portables were produced in large numbers. During World War II they were supplied to members of the Axis forces, for the Everest factory was one of the few in Europe still producing typewriters, most of the others having switched to the manufacture of war instruments and other munitions. After the war, production, sales, and exports of both their standard and portable machines again reached a high figure, but in 1962 the Company was taken over by Olivetti and they ceased producing typewriters of their own brand name. They lasted forty years.

THE MERCEDES: GERMANY

This was one of the best known and most popular German typewriters invented by Franz Schueler and improved upon by Karl Schlüns who had previously worked for fifteen years with Heinrich Kleyer, the manufacturer of the Adler typewriter. It was first manufactured in 1907, by the Mercedes Office Machine Company in Berlin, and a year later, the production was transferred to Zella-Mehlis in Thuringia.

The Mercedes was from the beginning a good solid Group 10 machine, with forward-strike, segment and bars, four rows of keys and a two-coloured ribbon. It had the advantage of interchangeable baskets for writing various languages, and also interchangeable carriages.

In 1921 Mercedes produced what was the first practical electric typewriter, and upon this design most modern electric typewriters are based. Two kinds of electric typewriters were produced, one a semi-electric and the other fully electric. The Mercedes 'Motor Express' had an electric carriage return, and on the Mercedes 'Electra' the bars were operated by electricity in the same manner as they are on most modern machines today.

For some time the Mercedes was sold in England under the name of 'Protos' and in Argen-

See pages 202–3

Table 19

PRODUCTION OF EVEREST TYPEWRITERS

Everest Portable

Model 90

Serial No.	Year
24,288	1937
27,930	1938
32,790	1939
38,381	1940
44,500 to 50,000	1941
Details unknown because of war events 1941–1943	
130,000	1944
132,610	1945
135,541	1946
140,721	1947
147,211	1948
154,401	1949
163,251	1950
174,377	1951
187,901 to 192,799	1952 (Jan. to July)

Model K.2

Serial No.	Year
500,001	1952 (July to Dec.)
505,021	1953
518,505	1954
530,851	1955
539,801	1956
554,191	1957
568,901	1958
589,301	1959
611,001 to 649,060 and 1,550,000 to 1,560,000	1960
1,560,427 to 1,581,200 and 1,590,001 to 1,600,110	1961
1,600,111 to 1,639,500	1962

Model K.3

Serial No.	Year
1,000,001	1960
1,001,201 to 1,025,600 and 1,050,001 to 1,061,484	1961
1,061,485 to 1,112,000	1962

Everest Standard Manual

Serial No.	Year
60,000	1937—1943
90,480	1944
95,833	1945
98,836	1946
104,738	1947
112,476 to 119,507	1948 (Jan. to June)

Model S.T.

Serial No.	Year
20,001	1948 (June to Dec.)
201,076	1948
209,901	1950
220,951	1951
234,651	1952
249,901	1953
262,031 to 267,595	end of production 1954

Model 92

Serial No.	Year
300,001	1954
303,101	1955
322,701	1956
340,620	1957
355,475	1958
369,801 to 384,499 and 384,601 to 385,300	1959
384,500 to 384,600 and 385,301 to 388,500 and 3,000,001 to 3,013,266 and	1960
3,700,001 to 3,700,910	
3,013,267 to 3,038,936 and 3,050,001 to 3,064,400 and	1961
3,700,911 to 3,720,739	

b

c

7.13 Mercedes Typewriter
a Mercedes-Elektra Model of 1921 made by Mercedes Office Machines, Germany. It was also known as the Drake-London, Kolumbus, Mercedes-Express, Mercedes-Favorit, Protos and Cosmopolita
b Standard Model of 1926 (Serial No. 143,020)
c Standard Model of 1940 (Serial No. 535,977)

tina as 'Cosmopolita'. Similarly, before World War I it had been intended to assemble the machine in England and sell it as the 'Drake/London'.

In December 1930, Underwood-Elliott Fisher took over the Company and the production of portables commenced. The preliminary assembly work was originally carried out in the factory of the Underwood Typewriter Company in the U.S.A. and the machines were then finished off in Zella-Mehlis, but after 1933, the portables were produced completely in the German factory. There is no need to describe the original Mercedes portable, as it was exactly the same as the four row Underwood portable.

Various models followed which were the simplified Model 33 with the single-coloured ribbon, and in 1934, the Model 34, called the Mercedes 'Prima'. Production of all machines appeared to cease in 1951 after a continuous history of forty-four years.

The Mercedes was a wonderful machine, of first-class design and manufacture, and the company had the distinction of producing the first practical electric typewriter from which all other models were copied.

THE OLIVER: ENGLAND

In 1888 the Rev. Thomas Oliver, a clergyman living in America designed and produced a typewriter for use in his work. By 1891, the first crude model of the Oliver typewriter was patented.

In December 1894 production and delivery of machines commenced with the delivery of Model No. '1', Serial No. 3, and this, by a strange coincidence, was delivered to another clergyman whose machine operated satisfactorily for over ten years before the owner was willing to exchange it for a later model.

By 1898 larger premises were occupied and the Oliver typewriter began to make its name in the world. Two years later, the Company were forced to make a further move to even larger premises, so enormous was the growth of the business.

The Oliver three-bank typewriter changed very little over the years and up to about 1935 appeared to have a ready market, being the only successful standard machine with three rows of keys.

During World War I, the Oliver was used extensively by the British Forces. It was strong and found to be reliable under almost any conditions, whether in the desert or on the Western Front.

In 1928 production was transferred to Croydon in Surrey, England, by which time more than one and a quarter million Oliver three-bank machines had been manufactured and sold. In Germany it appeared for a time under the name of 'Stolzenberg'.

In 1931 the three-bank machine was discontinued and various agents all over the world

Table 20

PRODUCTION OF OLIVER TYPEWRITERS

Oliver Manual	
3-Bank Models	
Nos. 5 and 6	1907—1914
Nos. 7 and 8	1914—1915
Nos. 9 and 10	1915—1922
Nos. 11 and 12	1923—1928
Nos. 15 and 16 below EA20,000	1938—1939
EA20,000	1940
EB24,751	1941
EC28,351	1942
ED31,801	1943
EE34,051	1944
EF34,152	1945
EG34,261	1946
EH34,346 up	1947
Model 20	
350,001	1935
360,001	1936
370,001	1937
380,001	1938
390,001	1939
400,001	1940
410,001	1941
420,001	1942
430,001	1943
440,001	1944
450,001	1945
460,001	1946
470,001	1947
480,001	1948
4,911,400	1949
5,000,001 upwards	1950
Model 21	
9,210,001 to 9,219,999	1949
0,210,000 to 0,219,999	1950
1,210,000 to 121-11149	1951
221-11500 to 221-15700	1952
321-15701 to 321-17279	1953
421-17280 to 421-18849	1954
521-18850 to 521-20299	1955
621-20300 to 621-21553	1956
721-21554 to 721-22428	1957
821-22429 to 821-23334	1958
921-23335 to 921-23490	1959
Oliver Portable	
Early Model	
To 70,000	1930
80,000	1931
83,501 and over	1932
Post-War Model	
4,800,001	1948
4,900,001	1949
5,000,001	1950
5,100,001	1951
5,200,001	1952
5,300,001	1953
5,400,001 to 5,450,000	1954
Courier	
540,001	1954
555,001 to 55-20,000	1955
56-10,000 to 56-40,000	1956
57-32,800 to 57-42,839	1947
A1/47,628 to A1/53,819	1957
BB/53,820 to BM/73,386	1958
CC/73,387 to CG/79,832	1959

were offered the German four-bank typewriter 'Fortuna' under the name of Oliver.

Four years later, in 1935, the production of the Halda-Norden standard typewriter under licence was produced as the Oliver Model '20'.

b

7.14 Early Oliver Typewriters
a Model 1 of 1894
b Model of about 1907. This was a three-bank machine belonging to Group 6 with the down-strike from the side. It changed very little from the beginning until it was discontinued in 1940

b

c

7.15 Oliver Typewriters
a Portable Model of 1930 (Serial No. 387,277)
b Portable Model of about 1930 belonging to Group 10 and a four-bank machine
c Standard Model 21 of 1951 (Serial No. 1,216,916). This was the Halda/Facit Typewriter made under licence in U.K.

At the beginning of World War II in 1939, the British Government placed a substantial order with the Oliver Typewriter Company for the three-bank Model '15' which had proved so reliable. As a result of this order the Company re-equipped their plant with their old machinery and supplied the Model '15' in 'war finish' for use by various units of the forces.

After World War II, in approximately 1947, the Oliver Company resumed production of the 'Halda-Norden', then called the Model '20', which was followed by a slightly modified version known as the '21'.

Simultaneously, production started on the 'Oliver' portable which was a well established basic design and sold in various countries as 'Fortuna', 'Augusta', 'Imperia', 'Littoria', 'Mas', 'Nuova-Levi', 'Sim', 'Byron', 'Japy', and 'Patria'. The name 'Patria', however, was protected by Swiss patents.

The Company lasted for some years but then discontinued production of the standard machine and concentrated on the portable, and decided to import from Germany the 'Siemag' Standard and sell this machine as the 'Oliver' standard. This venture was for some reason discontinued. At this point the Oliver Company appeared to be very successful. Fresh capital was introduced and they purchased in 1958 the Byron Typewriter Company Limited, previously the Barlock Typewriter Company, in Nottingham.

The Oliver Company then followed several courses of action, importing at various times not only the 'Siemag' Standard but also the 'Voss' from Germany, sold as the 'Oliver' portable, and the 'Japy Beaucourt' from France, sold as the 'Byron' portable, but all these ventures met with little success. They therefore concentrated on the production of portable typewriters in Croydon. After some time the machine tools for the small portables were transferred to the Voss factory in Germany to produce a machine which re-appeared in England as the 'Oliver'.

At this stage the Oliver Company stated they were going to become a Finance Company, and there were rumours of their merging with a vast group, but suddenly, about 1960, everything seemed to be out of control and years later in 1971, the shareholders received a letter from the Official Receiver saying that the affairs of the Oliver Company had not been resolved; what did emerge clearly was that shareholders and ordinary creditors would not receive any return. Thus ended another famous Company which had lasted about eighty years. This also marked the end of the Barlock and the Byron Typewriter Companies, so that at one fell swoop two British Typewriter Companies which had enjoyed many years of success disappeared.

THE SMITH PREMIER: U.S.A.

In 1889 the Smith Premier Typewriter Company commenced production of a typewriter which was invented by Alexander T. Brown; it had a seven-bank double keyboard and was an understrike machine. The carriage had to be lifted to see the typing.

Model '2', an improved version of the Model '1', was produced in 1895. It was followed by modified Models '3', '4', '5', '6', '7', '8', and lastly Model '9', which appeared in 1900.

In 1908 the Model '10' was introduced. This still had seven rows of keys but was a visible writing, forward front-strike machine with a conventional ribbon and a right-hand carriage return, and was totally different in design. The typewriter continued as the Smith Premier until 1921.

In 1922 its assets were absorbed by the Remington Typewriter Company which had been manufacturing a forward-strike four-bank machine for the Monarch Typewriter Company from 1904 to 1921. This latter company lost its identity in 1921 and for a short space of time the Monarch was marketed by the Remington Typewriter Company as the Smith Premier.

By 1923 the Smith Premier was fitted with segment bars and type guide and was known as the Model '50' and later, the Model '60', and controlled by the Remington Typewriter Company who allocated separate agencies and sold the machine independently. The Remington 'Noiseless' typewriter was also sold under the name of Smith Premier.

In 1940 the policy changed and after then the name Smith Premier was no longer used.

b

c

e

f

7.16 Smith Premier Typewriters
- **a Model 1 of 1888 with a double keyboard and an up-strike**
- **b Model 2 of 1895 with a double keyboard and an up-strike**
- **c Model 4 of 1900 (Serial No. 9,263) with a double keyboard and an under-strike**
- **d Model 10 of 1909 (Serial No. SL 60,147) with a double keyboard and forward-strike**
- **e Monarch Model of 1904–21**
- **f Model 30 (Monarch) of 1921 (Serial No. NK 20,410)**

Models '50' and '60' were discontinued, and the 'Noiseless' continued to be produced under the name of Remington only. So the Smith Premier in one form or another lasted for fifty-one years and then vanished.

THE STOEWER: GERMANY

In 1857 the German firm of Bernhardt Stoewer of Stettin-Grunhof, was well established as a manufacturer of sewing-machines and bicycles.

In 1903 they turned their attention to the production of typewriters and first produced a three-bank forward-strike machine which was extremely well made and quite successful.

In 1905 they produced a four-bank machine which had all the attributes of a modern well constructed typewriter. From then on the machine was continually improved.

In 1909 the firm produced the Stoewer 'Record' which was marketed in England as the 'Swift' and 'Swift Record'. In France it was known as the 'Baka V' and elsewhere as the 'Cito' and the 'Lloyd'.

In 1915 during World War I, experiments were in progress to produce the machine in England under the name of 'Barratt' and a few were either manufactured or assembled, but difficulties increased as the war continued, and the plan had to be abandoned.

A portable called the 'Elite' was also produced and the early models had three rows of keys.

a

b

7.17 Stoewer Typewriters
a Record Model of 1909 which was also marketed as the Swift and Conqueror
b Swift/Stoewer Standard Model of 1912 (Serial No. 30,795)

The factory in Stettin employed a staff of 2,500 to meet the demand for their production. They were one of the most prominent typewriter manufacturers on the Continent and were awarded seven gold medals for high-class workmanship.

In 1920 Lord Lascelles used some of his considerable fortune to finance the Conqueror Typewriter Company Limited which was to produce the 'Stoewer Record' in England, to be called the 'Conqueror'. The story of the failure of this venture is found in the section 'Typewriters in Use Today' in Chapter 2, page 89.

Manufacture of the Stoewer continued in Stettin until late 1929, by which time an improved standard and portable were being marketed. However, in that year, financial difficulties were encountered and in 1930, one part of the factory was sold to a Berlin firm to produce component parts and the other part closed down. A later model which was almost completed was never brought to the production line. The tools for the portable were purchased by Rheinmetall in Sommerda, who produced the machine under their own name.

The 'Stoewer', 'Swift' or 'Conqueror', or whatever you may care to call it, was a most unusual machine with a strange career. It was well designed and could have been one of the most important typewriters in history. Fate decided otherwise, however, and it never achieved the place in the world that it rightly deserved. The production lasted twenty-seven years.

THE TORPEDO: GERMANY

The Torpedo was a development of the 'Hassia' typewriter which was produced in 1904, by Hermann Wasem, though some authorities state that the machine started life in 1893. In any event, by 1907 the machine, a four-bank front-strike Group 10, was being manufactured by the Weil-Werke.

Subsequently, the manufacture was undertaken by the Torpedo-Werke, and the Model '5' was introduced in 1911, with an interchangeable carriage. The Company pursued a sound and

a

b

Early Torpedo Typewriters
a Early Model of unknown date
b Model of 1907

b

c

e

f

7.19 Torpedo Bluebird Typewriters
a Portable Model of 1935 (Serial No. 216,310)
b Portable Model of 1951 (Serial No. 544,322)
c Dynacord Model 18 non-tabulator of 1952. Model 18B had a tabulator
d Model 30 of 1952
e Standard Model of 1955 (Serial No. 804,404)
f Dynacord Standard Electric Model of 1961
g Dynacord Standard Model of 1965

Table 21

PRODUCTION OF TORPEDO, BLUEBIRD, AND DYNACORD TYPEWRITERS

1,101 to 1,500	
1,501 to 2,400	1908
2,401 to 6,500	1910
6,501 to 11,100	1911
15,001 to 15,400	1914
20,001 to 48,000	1913
48,001 to 72,250	1922
75,001 to 90,000	1925
90,001 to 95,000	1927
95,001 to 100,000	1928
100,001 to 110,000	
110,001 to 122,000	
122,001 to 132,000	1929
132,001 to 134,750	
134,751 to 145,000	
145,001 to 155,000	1931
155,001 to 170,000	
170,001 to 179,000	
179,001 to 201,000	
201,001 to 225,000	
225,001 to 227,000	1935
227,001 to 254,500	
254,501 to 287,000	
287,001 to 320,000	
320,001 to 355,000	
355,001 to 385,000	
385,001 to 405,000	
405,001 to 426,000	
426,001 to 440,000	
440,001 to 441,000	
441,001 to 442,000	

Model 2	1907
Model 3	1908
Model 4	1910
Model 5	1911
Model 5b	1913
Model Simplex	1914
Model 5c	1922
Model 12	1925
Model 6	1927
Model 14	1928
Modell 15	1931
Model 15a	1932
Model 16	1933
Model 17	1933
Model H15	1933
Model Su 6	1934

Standard Manual

To 218,000	1935
218,001	1936
240,001	1937
280,001	1938
290,001 to 300,000	1939
500,000	1950
541,001	1951
580,001	1952
650,001	1953
712,501	1954
775,001	1955
837,501	1956
900,001	1957
960,001	1958
1,030,001	1959
1,080,001 to 1,100,000	1960

Model 20—Ace Models

130,000	1931—1932
160,001	1932—1934
190,001	1935
220,001	1936
245,001	1937
277,001	1938
290,001 to 310,000	1939
500,000	1951
551,935	1952
650,001	1953
712,501	1954
775,001	1955
837,501	1956
900,001	1957
960,001	1958
Over 1,030,000	1959

Table 21 contd.

De Luxe Model

250,000	1938
300,001	1939
500,000	1950
541,001	1951
580,001	1952
650,001	1953
712,501	1954
775,001	1955
837,501	1956
900,001	1957
960,001	1958
1,030,001	1959
1,150,001	1960
1,207,001	1961
1,245,001	1962
1,271,001	1963
1,305,001 upwards	1964

Model 30

960,001	1958
1,030,001	1959
1,350,001	1960
1,400,001	1961
1,506,001 to 1,516,000	1962

Dynacords

2,000,000	1960
2,011,001	1961
2,026,001	1962
2,034,001	1963
2,053,001 upwards	1964

aggressive selling policy and rapidly grew into one of Germany's most important typewriter manufacturing Companies.

In 1914 a portable was introduced to meet the demand for lower prices. During the latter part of World War I, production of both portable and standard machines ceased for a time, but was resumed immediately afterwards; many were exported and sold under various names such as 'Regent', Bluebird', 'Harrod', etc. The 'Regent' portable was also exported in parts which were assembled in Leicester by the Imperial Typewriter Company, and sold by them as the 'Regent'. This arrangement did not last very long however, as the Imperial Typewriter Company imported the tools and made the machine in its entirety.

The first model off the production line in 1932 was presented to J. B. Priestley and with his permission was called the 'Good Companion' from the title of his famous novel which had then just been published. In the same year the Company was purchased by, and came under the control of, the Remington Typewriter Company.

During World War II, production of both standard and portable machines continued throughout, and at the end of the war, the company could count themselves fortunate that their factory near Frankfurt was in the Western Zone of Germany, for the factory was restored to its rightful owners and the production of typewriters continued.

The machine was sold for some time in England as the 'Bluebird' and later as the 'Dynacord'—both standard and portable. It was also marketed as the 'Remington' in certain areas. Production ceased in 1964, and the factory was disposed of by Remington Rand who transferred the portable plant to Holland where certain models of the Remington portables are produced today.

The Torpedo was a superb typewriter and was very much ahead of its time in design and appearance. Millions were manufactured and sold, but problems of finance, diversification, and intense competition finally led to the closure and dismantling of the plant.

Perhaps it would have been even more successful if it had not been given a name with such unhappy associations! It was one of the great machines and sold successfully and continuously for approximately fifty-seven years.

THE UNDERWOOD: U.S.A.

In 1895 Franz Xavier Wagner designed a typewriting machine which was the forerunner of the Underwood typewriter. He had designed many other models previously.

Between 1893 and 1895 he worked and devised a new and radically improved machine and was joined by Mr. W. F. Helmond in 1894. Mr. Helmond assisted him in his selection of the different sizes and qualities of steel most suitable for his purpose.

The model they had produced was shown to John T. Underwood who at that time was a manufacturer of carbon papers, inks, and typewriter ribbons, and who was very cross because he had lost the Remington contract to supply these. Remington told him they were going to make the ribbons themselves. Underwood recognized that here, for the first time, a machine had been developed on which the typist could see every letter as it was being typed. He saw the tremendous possibilities of this important achievement and thus supported the scheme.

On 29 March 1895 the Wagner Typewriter Company was launched, and the original typewriters were manufactured by Lambert and Edgar. However, Lambert and Edgar made only 500 machines before the need to expand was so great that they found new premises on Hudson Street, New York City, and the Wagner Typewriter Company began production themselves, later changing the name to the Underwood Typewriter Company.

The change to visible writing was a world-acclaimed accomplishment and sales increased so far beyond expectations that in May 1898, the manufacturing plant was moved to new premises at Bayonne, New Jersey. This plant had a weekly production figure of 200 machines and this, it was thought, would be more than enough to meet the sales demand. However, in 1901, more space was needed for the rapidly-expanding business and a new plant was established at Hartford, Connecticut.

The firm had a much needed boost in 1900 when the American Government gave them a contract for 250 Underwood typewriters. These were for use in the Navy, and Underwood used this order as a vital point in their advertising at this time—especially as the visible writing machine was still in its infancy.

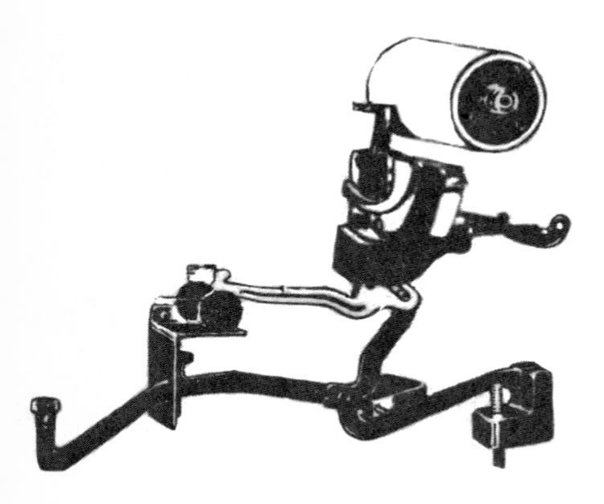

7.20 Shows view of Underwood manual action of 1900–31

7.21 Underwood Typewriters (see opposite)

a

b

d

e

f

g

7.21 Underwood Typewriters
- a Model 1 of 1896–7
- b Button Shift Model of 1906 (Serial No. 102,006) was of a single colour and had no backspace
- c Model 5 of 1915 (Serial No. 761,633)
- d Model 5 of 1926 (Serial No. 2,058,608)
- e Early Model 6 of 1934 (Serial No. 4,228,128)
- f Master Model of 1940 (Serial No. M 5,056,803)
- g Rhythm Shift Model of 1947 (Serial No. 6,217,757). This was the first segment shift machine produced by Underwood

a

b

7.22 Underwood Noiseless Typewriters
- a Portable Model of 1931
- b Model of 1936 (Serial No. 3,914,734)

The Hartford plant produced over 12,000 standard typewriters in the first year and eventually increased to hundreds of thousands annually. By 1939, five million Underwood machines had been produced and marketed in all parts of the world.

In 1926 Elliott Fisher and Sundstrand were brought together and they later merged with the Underwood Typewriter Company. In 1933 the manufacture of Elliott Fisher and Sundstrand products was consolidated in the works at Bridgeport, Connecticut.

In 1936 the Underwood Elliott Fisher Company formally opened its General Research Laboratory in Hartford. The purpose of establishing and maintaining this was to serve the interests of the businessmen of America and abroad. Talented engineers, scientists, chemists, technicians, machinists, and experienced business machine experts were housed under one roof in order to refine and develop their products, and to explore the possibilities of further expansion.

During World War II, Underwood concentrated chiefly on the manufacture of the U.S. carbine calibre ·30 M-1 producing 1,000,000 carbine barrels in the first fifteen months. In all,

7.23 Underwood Standa
Typewriters
a Model 150 of 1955
b Touchmaster II Mo
of 1960
c Touchmaster V Mo
of 1961
d Typemaster VI Mo
of 1966

7.24 Underwood Electri
Typewriters
a Model 1 of 1947
b Forum Model of 19
c Raphael Model of 1
d Documentor Mode
of 1961

Table 22

PRODUCTION OF UNDERWOOD TYPEWRITERS (NOW OLIVETTI)

Underwood

In 1929 the Underwood Company grouped their serial numbers and began serial numbering all models consecutively at 3,500,000

No. 3—11, 12, 14, and 16 in. Carriages

B/S and Bichrome began about	32,000
Key S/L began about	132,000
Rising Scale began about	455,000
Top Ribbon Switch began about	3,690,000
Under 28,500	1900–1908
28,501 to 128,000	1909–1914
128,001 to 400,000	1915–1922
400,001 to 455,000	1922–1923
455,001 to 600,000	1924–1926
600,001 to 780,000	1927–1928
780,001 to 3,500,000	1929
3,500,001 to 3,690,000	1929–1930
3,690,001 to 4,000,000	1930–1931

No. 3—18, 20 and 26 in. Carriages

B/S and Bichrome began about	6,000
Key S/L began about	28,000
Rising Scale began about	98,000
Top Ribbon Switch began about	3,690,000
Under 3,500	1903–1907
3,501 to 28,500	1908–1914
28,501 to 92,000	1915–1923
92,001 to 100,000	1924
100,001 to 145,000	1925–1927
145,001 to 3,500,000	1928–1929
3,500,001 to 3,690,000	1929–1930
3,690,001 to 4,000,000	1930–1931

No. 5—10 in. Carriage

B/S and Bichrome began about	245,000
Key S/L began about	760,000
Rising Scale began about	1,750,000
Top Ribbon Switch began about	3,690,000
To 37,500	1901–1903
37,501 to 247,000	1904–1908
247,001 to 760,000	1908–1914
760,001 to 1,610,000	1915–1922
1,610,001 to 1,750,000	1923
1,750,001 to 2,000,000	1924–1925
2,000,001 to 2,300,000	1925–1927
2,300,001 to 3,500,000	1928–1929
3,500,001 to 3,690,000	1929–1930
3,690,001 to 4,000,000	1930–1931

No. 6—Master Model—Rhythm Touch

Champion Keyboard began about	4,330,000
Rhythm Touch began about	6,070,000
4,000,000 to 4,010,000	1932
4,010,001 to 4,130,000	1933
4,130,001 to 4,265,000	1934
4,265,001 to 4,330,000	1935
4,330,001 to 4,400,000	1936
4,400,001 to 4,650,000	1937
4,650,001 to 4,800,000	1938
4,800,001 to 5,000,000	1939
5,000,001 to 5,260,000	1940
5,260,001 to 6,000,000	1941–1946
6,000,001 to 6,250,000	1947–1948
6,250,001 to 6,800,000	1949–1950
6,800,001 to 7,000,000	1951–1952
7,000,001 to 7,326,227	1953
7,326,228 to 7,447,999	1954

Model 150

7,570,000 to 7,660,103	1955
7,660,104 to 7,868,338	1956
7,868,339 to 7,975,836	1957

Touchmaster I

8,034,244 to 8,071,394	1957
8,071,395 to 8,124,559	1958
8,124,560 to 8,157,994	1959
8,157,995 to 8,290,000	1960

Touchmaster II

8,290,001 to 8,356,700	1960
8,356,701 to 8,512,311	1961

Touchmaster V

8,512,312 to 8,871,139	1961
8,871,140 to 9,133,814	1962
9,133,815 to 9,357,367	1963

Noiseless

To 3,800,000	1930–1933
3,800,001 to 3,900,000	1933–1934
3,900,001 to 3,930,000	1934–1935
3,930,001 to 3,950,000	1936
3,950,001 to 4,000,000	1937
4,000,001 to 4,400,000	1938
4,400,001 to 5,000,000	1939
5,000,001 to 5,230,000	1940
5,230,001 up	1941–1947

Underwood Electrics

SX-80

up to 6,200,000	1947

De Luxe

up to 6,524,499	1951

Finger-Flite

up to 7,275,000	1953

SX-153

up to 7,746,100	1956

SX-165

up to 7,908,000	1957

SX-167

up to 8,184,000	1959

Documentor, Forum, and Raphael

up to 8,593,624	1961
up to 8,871,404	1962
up to 9,134,149	1963
up to 9,356,859	1964
up to 9,540,827	1965

Noiseless Portable

500,000 to 602,000	1931–1933
602,001 to 607,000	1934
607,001 to 700,000	1935
700,001 to 736,000	1936
736,001 to 800,000	1937
800,001 to 900,000	1938
900,001 to 1,140,000	1939

Weight (in case) $16\frac{1}{2}$ lb. Dimensions of case: Height $5\frac{3}{4}''$, Width $12\frac{1}{2}''$, Length $12\frac{1}{4}''$.

Portable 3-Bank Model

Under 26,500	1920–1921
26,501 to 100,000	1922–1923

contd. overleaf

Table 22—cont.

Serial Numbers	Date
Portable 3 Bank Model—cont.	
100,001 up	1924–1929
Weight (in case) $9\frac{3}{4}$ lb. Dimensions of case: Height $5\frac{1}{2}''$, Width $10''$, Length $12\frac{1}{4}''$.	
Underwood Portable 4-Bank Models	
To 100,000	1926–1928
100,001 to 250,000	1928–1930
250,001 to 600,000	1930–1933
600,001 to 700,000	1933–1935
700,001 to 816,000	1935–1937
816,001 to 1,000,000	1937–1938
1,000,001 to 1,113,000	1939
1,113,001 to 1,250,000	1940
1,250,001 up	1941–1947
Weight (in case) $13\frac{1}{2}$ lb. Dimensions of case: Height $5\frac{1}{2}''$, Width $12''$, Length $12\frac{1}{8}''$.	
Model 18 Portable	
401,120 to 412,900	1960
412,901 to 432,000	1961
432,001 to 470,000	1962
470,001 to 510,000	1963
Model 21 Portable	
00,026 to 10,100	1961
10,101 to 18,000	1962

1,706,436 were made. They also made Rate-of-Climb Indicators for the U.S. Air Forces, bomb fuses, anti-tank components, and other ordnance items, as well as vital components in the B-29 and A-26 bomber programmes such as the automatic gun charger and writing devices used in their fire control system.

On 22 March 1945 the name of the Company was changed to the Underwood Corporation.

In order to secure much-needed financial development and integration of the production line, an agreement for close co-operation was reached with Olivetti of Italy in 1959. By 1960 Underwood had produced and sold over 12 million office machines and vast numbers of portables.

Underwood are now completely controlled by Olivetti and, for practical purposes, have lost their identity, although Olivetti sometimes make use of the Underwood brand name.

For sixty-five years Underwood produced one of the best and most popular machines, the design of which was fundamentally sound, and it varied only slightly throughout its entire history. In fact all modern machines are basically of this design.

a

b

7.25 Yost Typewriters

a Model of 1887 with a double keyboard and an understrike

b Model of 1908 was a four-bank visible writer with no ribbon and it operated from an ink pad

THE YOST TYPEWRITER: U.S.A.

The Yost typewriter was the invention of George Washington Newton Yost. The first machines were produced in 1887; they were non-visible with double keyboards and renowned for their excellent work. Yost was an inventor and an engineer who had been concerned with earlier types of writing machines and had gained practical experience with almost every inventor and manufacturer of typewriters up to that time. He laid down two important features: one, that the alignment should be satisfactorily maintained by means of a type guide and two, that ink should be applied by an ink pad.

By 1902 a machine had been designed which had a four-bank keyboard. It had many sizes of interchangeable carriages which ranged from 8½ to 36 in. and it was obtainable in all languages, including the Arabic and Persian languages in which the writing goes from right to left.

Because the Yost typewriter did not use a ribbon but an ink pad, it was outstanding for its clarity and sharpness of impression and it was found useful for producing offset litho plates.

The Yost Typewriter Company produced countless thousands of machines in the thirty-seven years of its life from 1887 until 1924, when production ceased. Full responsibility for the supply of spare parts was then accepted by the Remington Typewriter Company.

CHAPTER 8

A Compendium of Typewriter History

This chapter contains a list, arranged in alphabetical order, of most typewriters known to have been produced, together with quality photographs, in date order where possible; it includes brief illustrated details of typewriters which have not been dealt with separately in previous chapters.

Where there is a date against the photograph, this is believed to be the actual date of manufacture of the machine shown and is not necessarily the date of the first model. This latter date is shown in the compendium.

For the purposes of brevity, information against each photograph of the machine is given in the following order:

(*a*) Name of Machine
(*b*) Model of Machine
(*c*) The first known date of production
(*d*) Manufacturing Company
(*e*) The country (or countries) in which it was produced
(*f*) Other names (if any)
(*g*) Numerals 1–11 appearing in brackets indicate Group (*see* pages 221–3)
(*h*) Any other relevant information

The only means of acquiring accurate information is through manufacturers, museums, early volumes in the possession of the British Typewriter Museum, and kind persons who have passed on dates and information from memory. The position concerning dates, however, is confused by the fact that some books give different dates from others for certain machines. This may be because some typewriters were patented first and produced afterwards, while others were produced first and then patented. Examples of this are as follows:

The 'Lambert' which dates back to 1883, took seventeen years to develop and was mentioned as being produced both in 1883 and in 1896. Further confusion is probably caused because the machine was made by 'The Gramophone and Typewriter Company' in England, in Germany by the German Gramophone Company, and in France, in Dieppe.

Another clear example is the 'D.M.G.' made by the Daimler-Benz Motor Company,

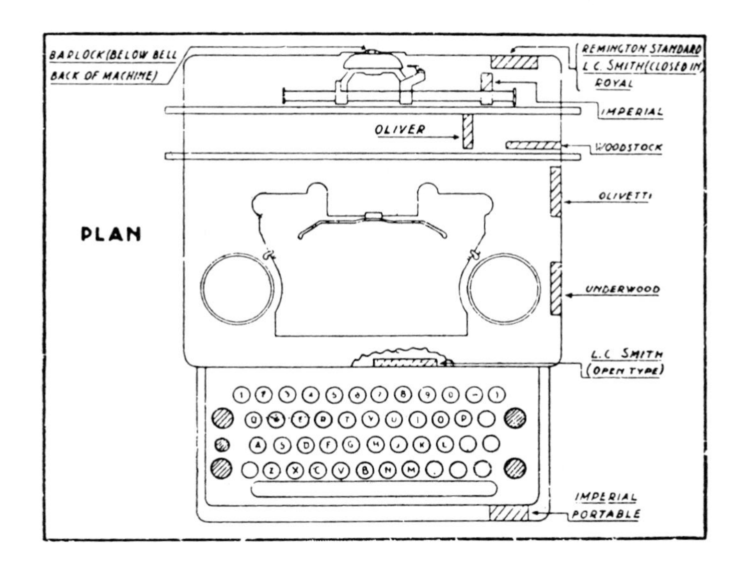

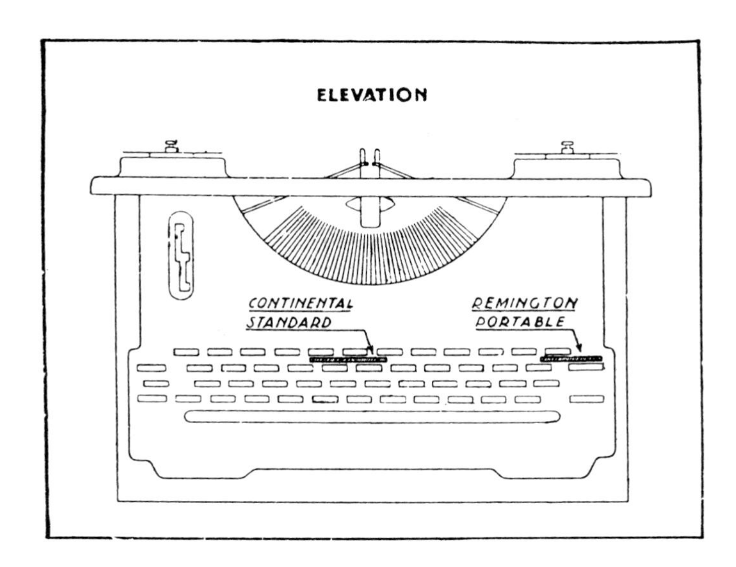

8.1 Location of Serial Numbers

who made a few machines only and then discontinued production. The patents were not granted, however, until some years later.

A further example is the 'Hall', which was patented in 1867, but production did not start until 1881. Most dates vary by one or two years only, however.

It would obviously be quite impracticable to deal with the many thousands of obscure brand names which still are well-known makes in disguise, though some of these are included.

Mention should be made of Serial Numbers; these are put in different places on the various makes of machines and Figure 8.1 is a plan showing the location of these numbers.

Table 23

SUMMARY OF TYPEWRITER CATEGORIES

Most typewriters fall into one of the following twelve groups (*see* Figures 8.2 to 8.12, pages 223–5)

Group 1 Swinging Sector Designs
e.g. Hammond Multiplex

Group 2 Type-Wheel Designs
e.g. Blickensderfer; I.B.M. Golf Ball '72/82'

Group 3 Type-Sleeve Designs
e.g. Crandall

Group 4 Radial-Strike Plunger Designs
e.g. Hansen machine; 'Noiseless' typewriters

Group 5 Index System Designs
e.g. Mignon

Group 6 Bar Type Designs
Up-strike: e.g. Remington No. 1
Down-strike (three kinds):
from Front;
from Side;
from Rear.

Group 7 Grasshopper Movements
e.g. Williams

Group 8 Semi-Front-Strike Designs
e.g. Based on the Corona

Group 9 Front-Strike Design
with bearings side by side or staggered.
e.g. L. C. Smith Bros; Remington 10

Group 10 Front-Strike Design
with type guide and slotted segment; either carriage or segment shift for capitals.
e.g. All modern and currently produced machines, portable, standard and most electrics fall within this group

Group 11 Electric Typewriters
All current models, with the exception of the IBM '72/82' are front-strike with typebars and slotted segments. As with standard office typewriters, the segment shifts upwards or downwards for capital letters with certain exceptions where the carriage moves up and down.

Some manufacturers have produced machines where the carriage is manually controlled.

Table 23 contd.

Group 12 Not illustrated.
This is the one exception in the electric typewriter field. The IBM '72/82' is basically an ingenious modification of Group 2 Type-Wheel Design similar to Blickensderfer but differing from all other machines in that the carriage remains stationary and the type ball head moves along the rubber cylinder.

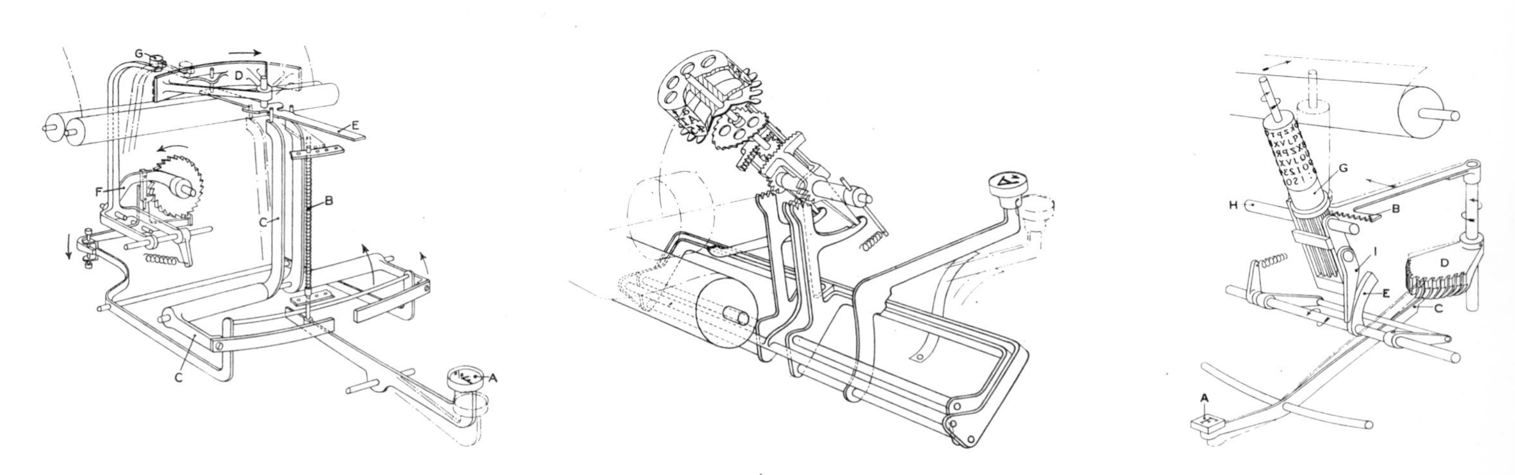

8.2 Group 1: Swinging-Sector Mechanism (based on the Hammond machine)

8.3 Group 2: Type-Wheel Designs, e.g. Blickensderfer, IBM Golf Ball 72/82

8.4 Group 3: Type-Sleeve Designs, e.g. Crandall (mechanism based on the Crandall machine)

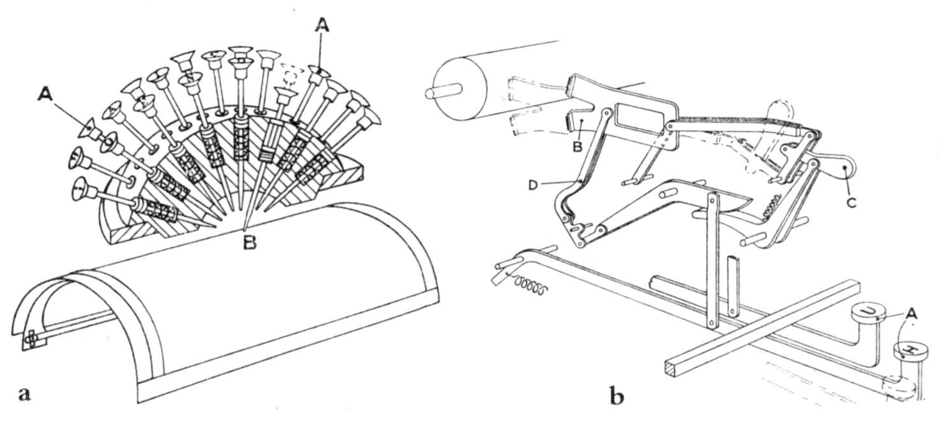

8.5 Group 4: Radial-Strike Plunger Designs e.g. Hansen Machine, 'Noiseless' Typewriters
a View based on the Hansen machine. The radial key plungers AA all strike to a common point B on the surface of the platen
b View of Kidder's principle, e.g. Adler, Wellington

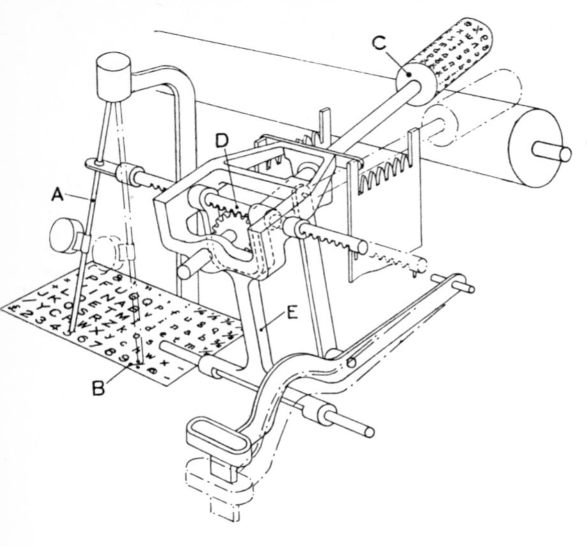

8.6 Group 5: Index System Designs, e.g. Mignon

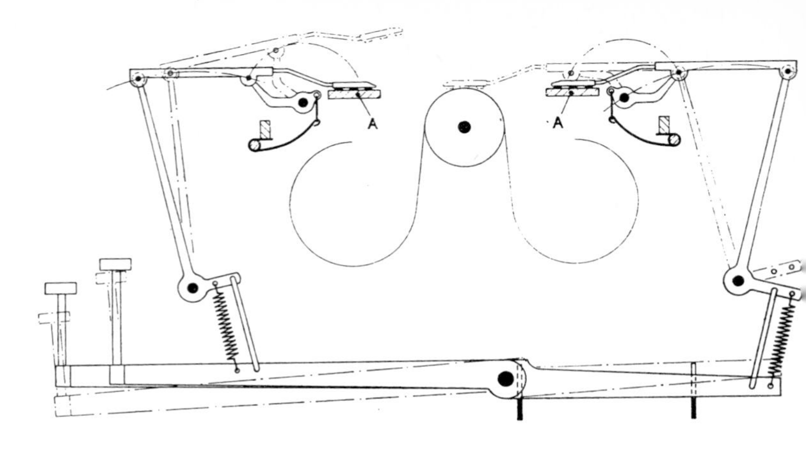

8.8 Group 7: Grasshopper Movements, e.g. Williams

8.7 Group 6: Bar Type Designs

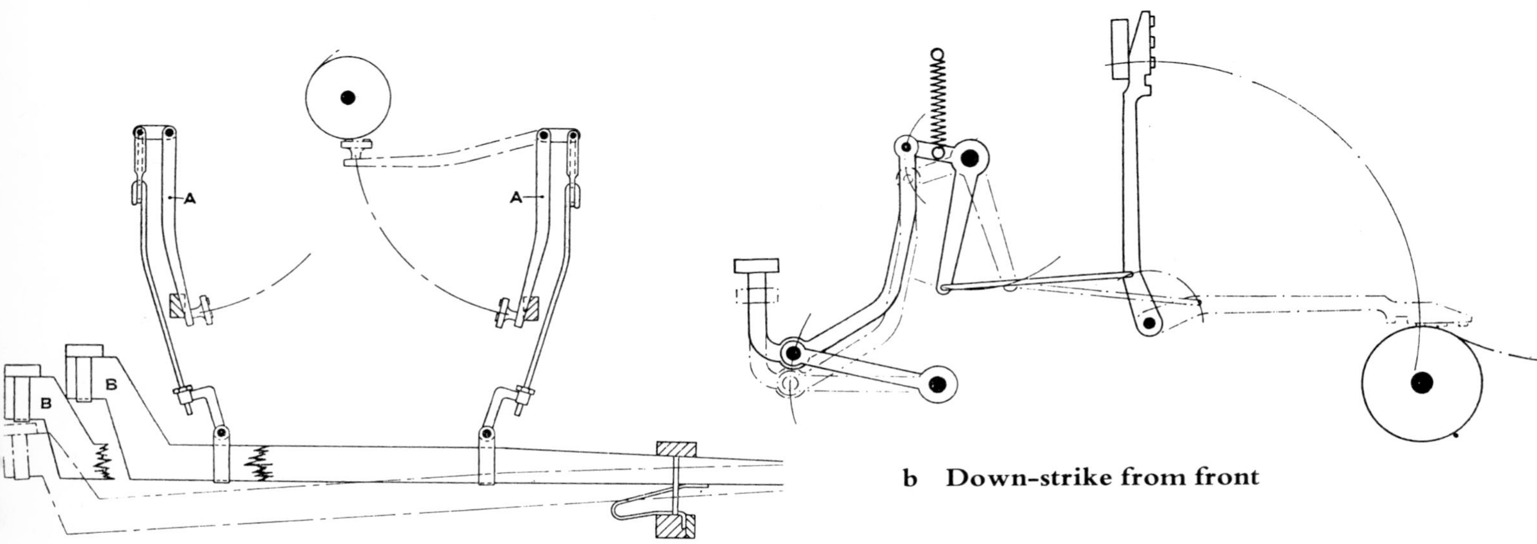

a Up-strike, e.g. Remington No. 1

b Down-strike from front

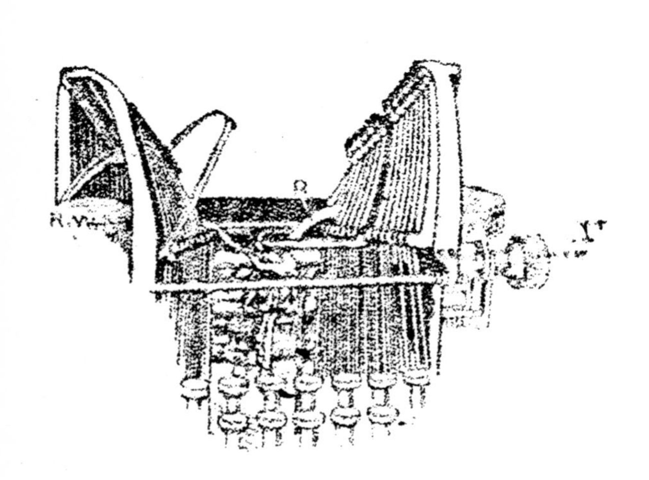

c Down-strike from side

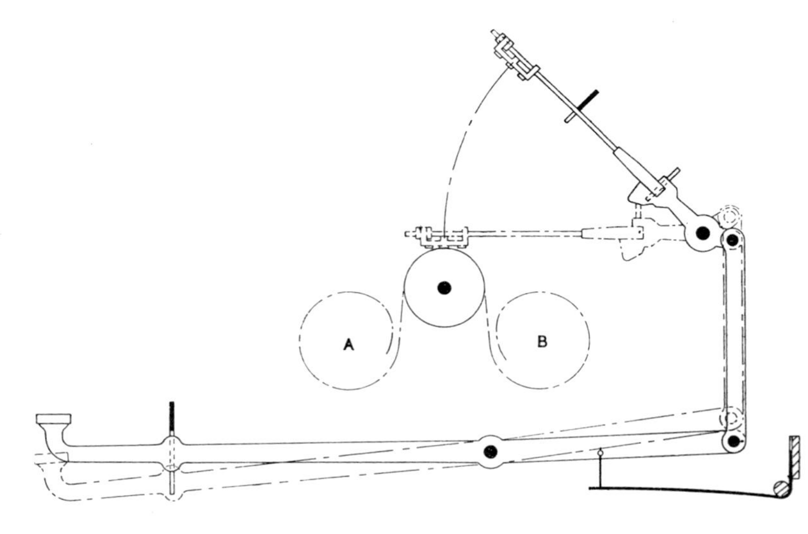

d Down-strike from rear (based on the Fitch machine); another example is North's Machine

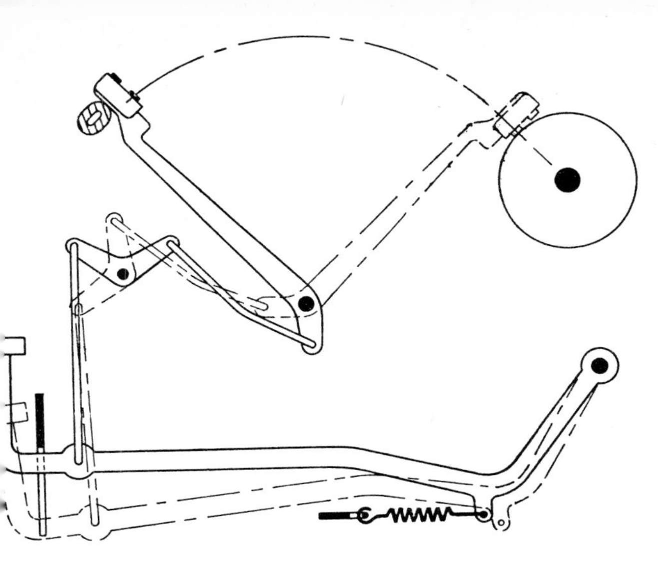

8.9 Group 8: Semi-front-strike Designs, e.g. based on the Corona

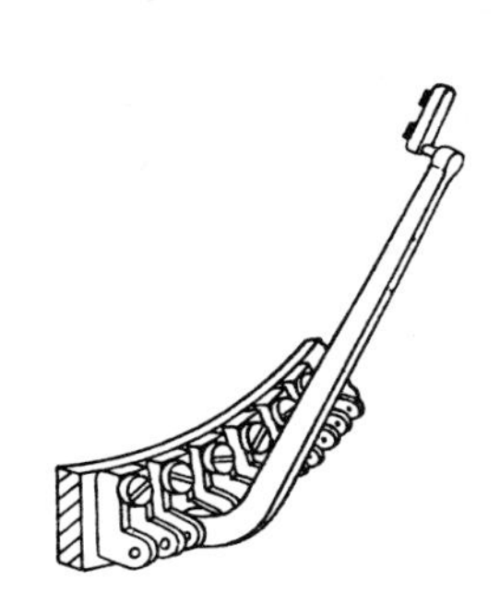

a Bearings side by side

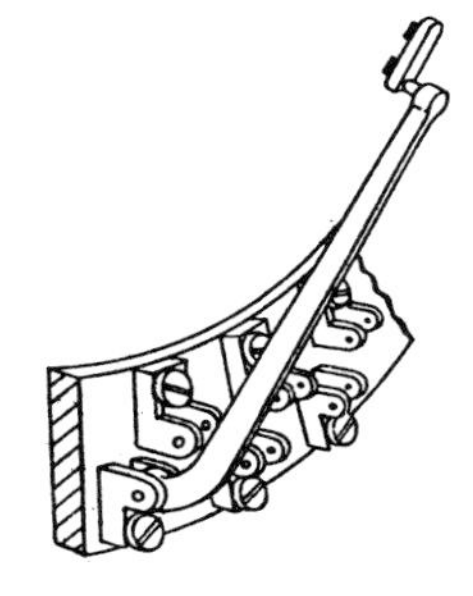

b Bearings staggered

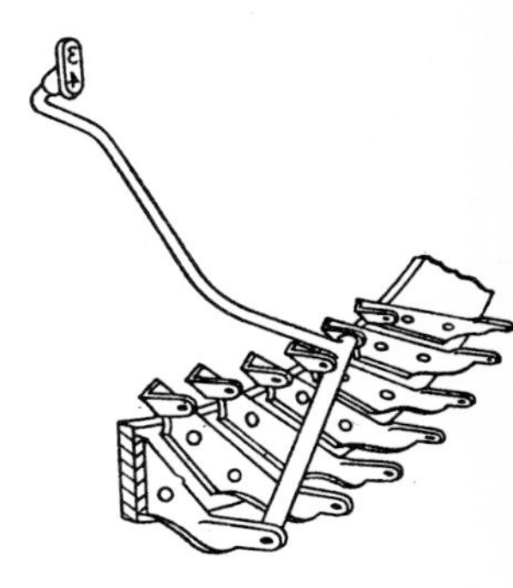

c Bearings oblique

8.10 Group 9: Front-strike Design, e.g. L. C. Smith Bros, Remington 10

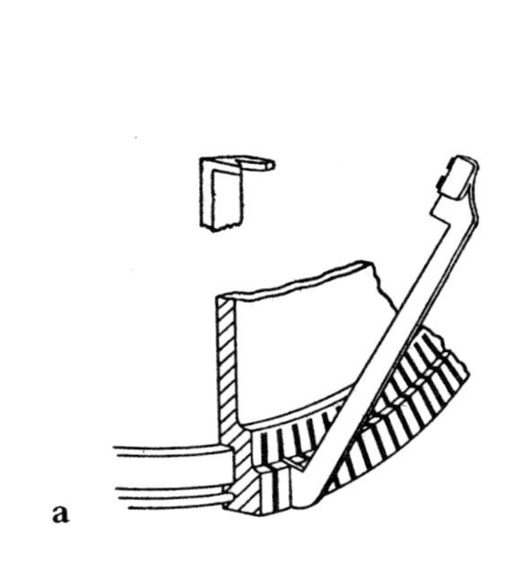

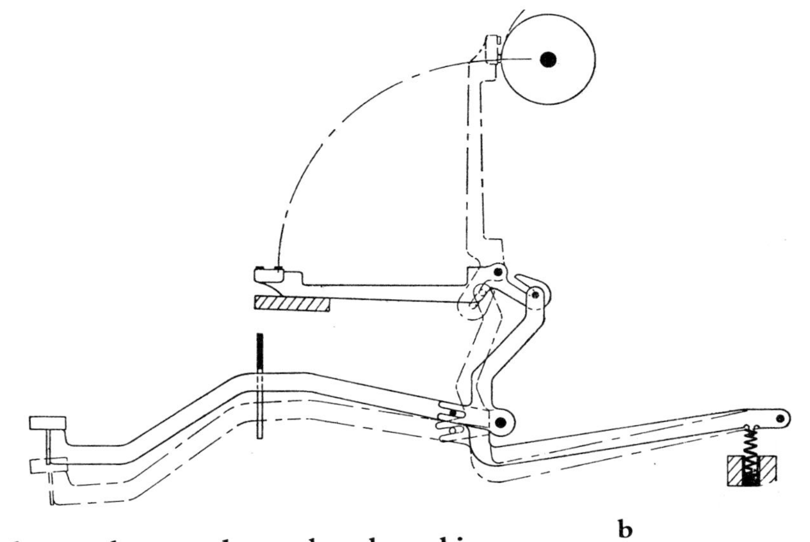

.11 Group 10: Front-strike Design, e.g. all modern and currently produced machines, portable, standard and most electrics fall within this Group

a Slotted guide-comb. Pressed sheet metal typebars with type guide

b Front-Strike mechanism (based on the Underwood machine)

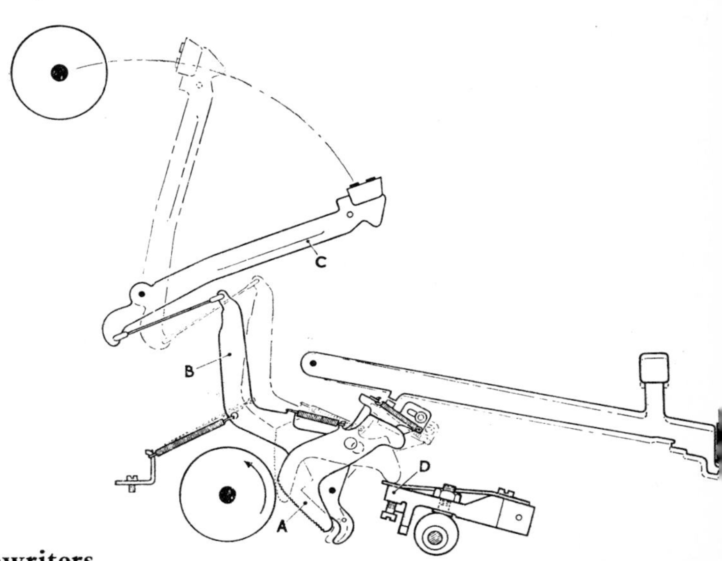

8.12 Group 11: Electric Typewriters

Pastor Malling Hansen's 'Writing Ball'

*Machines marked with an asterisk will also be found in the accompanying illustrations, Figures 8.13–8.179, between pages 246 and 265

Name	*Electric, Standard or Portable*	*Country of Origin*	*Date*	*Other Names if any; Remarks*
A				
Ace	P	U.S.A.	1934	Underwood
Acme	S	U.S.A.	1911	
Adler	S	Germany	1898	*See* Chap. 3 (Adler)
Adler	P	Germany	1912	Adlerette. Aigle. Adlerita. Adler Piccola
A.E.G.	S	Germany	1921	Olympia. Mignon
A.G.	S	Germany	1921	Olympia. Mignon
Agamli-Ogli	S	Germany	1929	Yanalif
Agar	P	Italy	1922	Juventa
Agar Baby	P	Italy	1922	Juventa
Aigle	P	Germany	1912	Adler
Ajax	Junior	U.K.	1908	Imperial
Aktiv	S	Germany	1913	Famos
Al Ahram	S	Germany	1920	Rheinmetall
Ala	P	Switzerland	1935	Hermes Baby
Alba	P	Italy	1954	Antares
Albus	P	Austria	1909	
*Alexander	S	U.S.A.	1907	
Alfa	S	Germany	1910	Minerva
*Allen	P	U.S.A.	1918	
Allen Electrite	S	U.S.A.	1955	*See* Chap. 7 (Woodstock)
Allen R. C.	S	U.S.A.	1950	*See* Chap. 7 (Woodstock)
All New (Personal)	P	U.S.A.	1951	Remington Portable
Alpina	P	Germany	1951	
Amata	S	Austria	1923	
Ambassador	S or E	Switzerland	1953	Hermes
A.M.C.	P	Germany	1922	Orga-Privat. Nagri
*American		U.S.A.	1899	Daugherty
*American Pocket	P	U.S.A.	1926	
*American Standard	S	U.S.A.	1892	Jewett
*American Visible	P	U.S.A.	1893	
Amka	P	Germany	1926	Bing
Anavi	S	Austria	1924	
Annell	S	U.S.A.	1922	Woodstock: R. C. Allen
Antares	P	Italy	1954	Mercedes (Italian) Olivetti
Archo	S	Germany	1920	
Ardita	P	Germany	1922	Juventa, later Everest
Aristocrat	P	U.K.	1948	Empire (Hermes)

Name	Electric, Standard or Portable	Country of Origin	Date	Other Names if any; Remarks
Aristokrat	P	Germany	1935	Urania
Arlington	S	U.S.A.	1914	
*Armstrong	S	U.K.	1907	American
*Arnold	Index	U.S.A.	1876	
Arpha	S	Germany	1921	Commercial
Artus	S	Germany	1912	Minerva
Astoria	P	Italy	1936	Oliver (Italian)
Atlantia	S	Germany	1921	Commercia. Montana
Atlantic	P		1948	Experimental only
Atlas	P	U.S.A.	1915	
Augusta	P	Italy	1930	Juventa. Oliver (Italian)
Aurora	S	Italy	1934	Combination of ordinary typewriter and Blind machine
*Autocrat	S	U.S.A.	1911	Harris
*Automatic	P	U.S.A.	1884	
Aviso	Junior	Germany	1923	
Avona		Germany		
Azzurra	P	Switzerland	1951	Hermes (in Italy)
B				
Baby	P	Switzerland	1951	Hermes
Baka I	P	U.K.	1907	Moya (in France)
Baka V	S	Germany	1903	Stoewer (in France)
Balkan	P	Germany	1913	Senta
Baltika	S	Denmark	1928	
Baltimore	S	U.S.A.	1890	Munson
Bamberger	P	Germany	1908	
Bantam	P	U.S.A.		Remington
Bar-Special	P	U.K.	1926	
Barlet	P	U.K.	1931	Mitex. Tell
Bar-lock	S	U.K.	1889/1918	*See* Chap. 7
*Barr-Universal	P	U.S.A.	1926	Barr. Morse
Barratt	S	U.K.	1915	Stoewer. Swift. Conqueror. *See* Chap. 7
Bavaria	Junior	Germany	1921	
Beaucourt	P	France	1959	Japy. Oliver
Beko	P	Germany	1926	Bing
Belka	S	Germany	1921	Reliable
*Bennett	P	U.S.A.	1910	Junior
Bennington	P	U.S.A.	1903	
Berni	P	Germany	1926	Bing
Berolina	Junior	Germany	1919	Meteor. Nova. Commercial

Name	*Electric, Standard or Portable*	*Country of Origin*	*Date*	*Other Names if any; Remarks*
*Bijou	P	Germany	1910	Erika. Gloria
*Bing	P	Germany	1925/7	Amka. Berni. Beko
Blake	P	U.S.A.	1905	Boston and World
Blick				*See* Chap. 7
Blick-Bar	S	U.S.A.	1916	Moyer
Blick-Ninety	P	U.S.A.	1919	
Blickensderfer	P	U.S.A.	1889	Dactyle
Blickensderfer Electric	S, E	U.S.A.	1902	
Bluebird	P	Germany	1931	Torpedo
Borgo	S	Germany	1922	Minerva
*Boston	P	U.S.A.	1886	World. Blake
*Brady-Warner		U.S.A.	1878	Experimental only
Brandenburg	S	Germany	1911	Franconia (Excelsior)
British Empire	S	U.K.	1924	
British Empire	P	Canada	1895	Kidder's Machine
British Imperial	S	U.K.	1922	Imperial
British Oliver	S	U.K.	1928	Halda (4-Bank)
Broadway Standard	S	U.S.A.	1898	Pittsburgh
*Brooks	S	U.S.A.	1885	
Brosette	P	Germany	1953	
Brother	P	Japan	1964	Remington in U.S.A., other countries
Brother	Compact	Japan	1969	Electric. Remington, other countries
*Burnett	S	U.S.A.	1908	Triumph. Perfect. Visible
*Burns	S	U.S.A.	1889	
Burroughs	S	U.S.A.	1931	
Burroughs	E	U.S.A.	1932	
Byron	S	U.K.	1952	*See* Chap. 7 (Barlock)
Byron	P	France	1959	Japy. Byron (Portable). Beaucourt

C

Name	Type	Country	Date	Other Names
Cadet	P	U.S.A.	1939	Remington
Calanda	P	Switzerland	1942	
*Caligraph	S (Blind)	U.S.A.	1881	Century
Cappel	S	Germany	1914	Kappel
Carat	S	Germany		Rheinmetall
Cardinal	S	Germany	1923	
*Carissima	S	Germany	1934	
Carlem	S	U.S.A.	1885	Sun
Carlem	P	U.S.A.	1913	Sun
Carmen	Junior	Germany	1920	

Name	*Electric, Standard or Portable*	*Country of Origin*	*Date*	*Other Names if any; Remarks*
★Cash	S	U.S.A.	1887	
Celtic	S	France	1923	Herkunft
Celtic	P	France	1923	
★Century	S	U.S.A.	1919	Remington Junior (1914–18)
★Champion	S	U.S.A.	1893	Underwood—Peoples
Chancellor	S	Germany	1903	Kanzler
★Chicago	Junior	U.S.A.	1898	Munson
Cicero		Sweden		Halda
Cito	S	Germany	1904	Stoewer. Swift. Conqueror
Cleveland	S	U.S.A.	1894	Hartford
Clipper	P	U.S.A.		Smith-Corona
★Coffman	Index	U.S.A.	1902	
Collegiate	P	Germany		A.B.C.
★Columbia Barlock	S	U.S.A.	1900	Barlock
Combina		Germany		Groma
Comet	P	Holland	1952	Smith-Corona
Commercial	P	U.S.A.	1901	
Commercial	S	Germany	1914	
★Commercial Visible	S	U.S.A.	1898	Fountain
Commodore	S	E. Germany	1961	Rheinmetall
Condor	S	Germany	1914	Commercial (German)
Concord	S	Germany	1926	Merz
Concordia	S	Germany	1908	Dea—Urania
Conover		U.S.A.		Munson
Constanta	S	Germany	1914	Commercial (German)
Consul	P	Czechoslovakia	1952	Diplomat. Zeta
Contal				Continental
Contenta				Continental
Contex				Passat
Contin	S	France	1922	Continental
Contina				Continental
Continental	S	Germany	1904	
Continental-Electric				Continental
Continental-Polyglott				Continental
Continental-Portable	P	Germany	1929	Contine. Continent. Condinal. Continenta. Conta
Continental-Schreiber				Continental
Continental Silenta	S	Germany	1934	Wanderer
Continental/W.A.G.				Continental
Continento				Continental
Continette				Continental
Cooper	P	U.S.A.	1856	
Corona	P	U.S.A.	1912	Smith-Corona
Coronet	P	U.S.A.		Corona 4

Name	*Electric, Standard or Portable*	*Country of Origin*	*Date*	*Other Names if any; Remarks*
Correspondent	S	Holland	1921	Rofa (Germany)
Correspondent	P			Underwood
Cosmopolitan	S	Germany	1921	Corona Mercedes
Courier	S	Austria	1903	Also name of Oliver Portable (England) 1950
*Crandall	S	U.S.A.	1879	
*Crary	P	U.S.A.	1892	
*Crown	S	U.S.A.	1887	
Culema	S	Germany	1920	
D				
Dactygam	Special	France	1920	
Dactyle	P	France		Blickensderfer
Dankers	P	Germany	1935	Flott. Juwel
Danuvia	P	Hungary	1921	
Darling	Indicator	U.S.A.	1910	
*Daugherty	Semi Std.	U.S.A.	1891	First Front-strike machine
Davis	S	Canada	1912	Adler. Empire. Wellington. Kidder
*Daw and Tate [Tait]	S	U.K.	1884	Only two were made
*Dayton	P	U.S.A.	1924	
Dea	S	Germany	1908	Union. Concordia. Radio
*Decker-Beachler		U.S.A.	1917	Pittsburgh—Reliance. Reliance—Premier. American
Deesse	P	Germany	1912	Stoewer. Elite
*Defi	P	U.S.A.	1908	Eagle
*Demountable	S	U.S.A.	1921	
*Densmore	S	U.S.A.	1891	Harris
Deutschland	S	Germany	1909	Minerva. Underwood
Diadema	P	Italy	1922	
Diamant	P	Germany	1922	
*Diamond	S	Germany	1923	
Diana	P	Germany		Royal
*Dictatype	Shorthand	U.S.A.	1928	Anderson (1886)
Diktator				Victor
Diplomat	P	Czechoslovakia	1950	Consul. Zeta
Diskreet	Indicator	Germany	1899	
D.M.	S	Germany	1930	Olympia
*D.M.G.	S	Germany	early 1920s	Discontinued 1925
Dolfus		France		
Dollar	P	U.S.A.	1890	
Domus	P	Italy		Antares

Name	*Electric, Standard or Portable*	*Country of Origin*	*Date*	*Other Names if any; Remarks*
Doropa	P	Germany	1911	Meteor
Drake	P, S	Germany		Mercedes
Draper	Junior	U.S.A.	1890	Munson. Chicago
Drugj	S	Germany	1921	Reliable
D.S.	P	Germany	1912	Stoewer. Elite
Dual-Rite		U.S.A.		Remington
Duo	S	Germany	1921	Rheinmetall
*Duplex	S	U.S.A.	1895	
Durable	P	Germany	1909	Triumph
D.W.F.	S	Germany	1923	
Dynacord	S	Germany	1965	Bluebird. Torpedo (1965–67)
Dynacord	P	Germany	1945	*See* Chap. 7 (Torpedo)
E				
Eagle	P	U.S.A.	1902	*See* Chap. 3 (Adler)
*Eckels	S	U.S.A.	1892	
*Eclipse		U.S.A.	1885	Similar to Brooks
Eclipse	Indicator	U.S.A.	1918	Yu Ess. Mignon. Olympia
*Edelmann	Indicator	Germany	1897	
*Edison	S	U.S.A.	1894	
Edita	Indicator	Germany		Triumph
Edland	Indicator	U.S.A.	1891	
Electa		Italy		Invictor
Electromatic	S	U.S.A.	1930	I.B.M.
Elgin	P	U.S.A.	1908	American. Eagle. Europa
Elite	P	Germany	1950	
*Elliott-Fisher	S	U.S.A.	1903	Underwood
*Elliott-Hatch	S	U.S.A.	1897	
Ellis	S	U.S.A.	1910	
Emerson	S	U.S.A.	1907	Later became Woodstock
Emka	P	Austria	1910	Albus
Emona	S, P	Yugoslavia	1952	
Empire	S	Canada	1892	Kidder
Empire	S	U.S.A.	1895	Kidder
Empire	P	U.K.	1960	Smith-Corona
Engler	P	Austria	1910	Albus
*English	S	U.K.	1890	
Envoy	P			Remington
Erika 3 Bank	P	E. Germany	1910	Ideal. Bijou
Erika 5	P	E. Germany	1927	Ideal. Bijou
Erika Late Models	P	E. Germany	1970	Sold U.K. as 'Boots'
Esko				Commercial
Essex	S	U.S.A.	1890	

Name	Electric, Standard or Portable	Country of Origin	Date	Other Names if any; Remarks
Eureka	Junior	U.S.A.	1896	
Europa	P	U.S.A.	1908	American. Eagle. Elgin
Everest	S	Italy	1937	Juventa. Saab. *See* Chap. 7 (Everest)
Everest Portable	P	Italy	1938	*See* Chap. 7 (Everest)
Everest 92	S	Italy	1955	*See* Chap. 7 (Everest)
Everlux	P	Italy		Antares
Excelsior	S	Germany	1921	
Excelsior Script	Script	U.S.A.	1899	Later Franconia, Cardinal
Express	P	Germany	1952	National
F				
Facit	P	Sweden		*See* Chap. 3 (Facit)
Facit	S	Sweden		*See* Chap. 3 (Facit)
Faktotum	S	Germany	1912	Forte-type
Famos	Indicator	Germany	1910	
Favorit	P	Germany		Adler American. Mercedes
★Fay-Sholes	S	U.S.A.	1901	
Featheright	P	Switzerland	1951	Hermes Baby
Federal	S	U.S.A.	1919	Visigraph
Felio	Junior	Holland	1919	Nora
Fidat	S	Germany	1912	Juventa. Minerva
Filius				Jewel
★Fitch	S	U.S.A.	1886	
Fiver	S	U.S.A.	1907	Oliver
Fix				Juwel. Dankers
F.K.	S	France	1932	
F.K. Mala	P		1938	Hermes Baby
Fleet	S	U.S.A.	1908	American
Flott		Germany	1935	Juwel. Dankers
F.N.	S	France	1921	MAP
Fontana	S	Italy	1921	Forte-type
★Ford	P	U.S.A.	1895	Forte-type
Forte-Type	S	Germany	1912	
Fortuna	S	Germany	1923	Oliver. Herkunft
Forum	S, P	Czechoslovakia		Zeta. Underwood Elect.
Fountain	P	U.S.A.	1898	
★Fox	S	U.S.A.	1906	
★Fox Baby	P	U.S.A.	1918	
★Fox 'Blind'	S	U.S.A.	1898	
★Fox Sterling	P	U.S.A.	1921	Fox. Rapid
Fox Sterling	S	U.S.A.	1906	Rapid. Fox-Visible
Framo	S	Germany	1914	Commercial
Franconia	P	Germany	1911	Standard folding
★Franklin	Junior	U.S.A.	1889	

Name	*Electric, Standard or Portable*	*Country of Origin*	*Date*	*Other Names if any; Remarks*
*Frister & Rossmann	S	Germany	1892	
*Frolio	P	Germany	1924	Gundka. Gefro
G				
Gabriele	P	Germany		Triumph. Adler
Galesburg	Junior	U.S.A.	1890	Munson. Chicago
Gallia	S	Germany		Urania
Galliette	P	France	1911	Perkeo
*Garbell	P	U.S.A.	1919	
*Gardner	P	U.K.	1893	
Gefro	P	Germany	1924	Frolio. Gundka
Geka	Indicator	Germany	1924	Geniatus
*Geniatus	Indicator	Germany	1928	Geka
Gerda	Blind	Germany	1919	
Germania	S	U.S.A.	1898	Germania-Jewett when sold in Germany
Gisela	P	Germany	1921	
Glashütte	S	Germany	1921	
Globe	S	U.S.A.	1893	American
Gloria	P	Germany	1923	Erika. Ideal. Bijou. Geniatus
Godrej	S	India	1955	*See* Chap. 3 (Godrej)
Good Companion	P	U.K.	1932	Imperial. Regent. Mead. *See* Chap. 4 (Imperial)
Gossen Tippa	P	Germany	1948	Afterwards: Adler-Triumph
*Gourland	P	U.S.A.	1920	
*Granville	S	U.S.A.	1896	
*Graphic	Indicator	Germany	1895	
Groma	S	Germany	1924	
Groma	P	Germany	1938	Kolibri
Gromina	P	Germany	1950	
Guhl	S	U.S.A.	1924	Stearns
Gundka	P	Germany	1924	Frolio. Gefro
H				
Hacabo	S	Germany	1912	Minerva
Haddad	S	Germany	1910	Urania
Halberg	P	Holland	1952	
Halda	S	Sweden	1896	*See* Chap. 3 (Facit)
Halda Portable	P	Sweden	1951	Facit. *See* Chap. 3
Halda Norden	S	Sweden	1930	Facit. *See* Chap. 3
*Hall	P	U.S.A.	1881	
*Hall Braille Writer	P	U.S.A.	1891	

Name	*Electric, Standard or Portable*	*Country of Origin*	*Date*	*Other Names if any; Remarks*
Hamilton	P	U.S.A.	1883	Automatic
Hammond	S	U.S.A.	1881	Later VariTyper
*Hammonia	Indicator	Germany	1883	
Hansa	S	Germany	1903	Kanzler
Hansa	S	Germany	1914	Commercial
Hansa	S	Germany	1919	Culema
*Hanson	S	U.S.A.	1899	(Experimental) Rex-Demountable
*Harris	S	U.S.A.	1911	Minerva. Autocrat. Reporter. Rex
Harrods	P	Italy		*See* Olivetti. Also Torpedo. Invicta
Harry A. Smith	S	U.S.A.	1917	Blick-Bar
*Hartford	S	U.S.A.	1894	Cleveland
Hassia	S	Germany	1904	Early Torpedo
Heady	Indicator	Germany	1921	Yu Ess. Mignon
Hebronia	S	Germany	1912	Minerva
Hega	P	Germany	1924	Phönix
Helios	P	Germany	1908	
Helma	P	Germany	1927	
Herald	P	U.S.A.	1908	
Hercules	S	Germany	1936	Bluebird. Torpedo
Hermes	S	Switzerland	1921	*See* Chap. 3 (Hermes)
Hermes Ambassador	S	Switzerland	1948	*See* Chap. 3 (Hermes)
Hermes Portable	P	Switzerland	1935	*See* Chap. 3 (Hermes)
Hermes 2000	P	Switzerland	1940	*See* Chap. 3 (Hermes)
Heroine	S	Germany	1923	Minerva. Reliable
Herold	S	Germany	1934	Rheinmetall
Heros	S	Germany	1914	Commercial
Hesperia	S	Italy	1921	Fontana
Hispano	S, P	Italy		Olivetti
*Hooven	S	U.S.A.	1912	
*Horton	S	Canada	1883	Hooven Automatic
Hurtu	P	U.S.A.	1895	Ford (U.S.A.) but sold as Hurtu in France
I				
IBM	S	U.S.A.	1933	International. *See* Chap. 4 (IBM)
IBM Electric	S	U.S.A.	1946	
Idea	S	Poland	1923	Pacior
Ideal	S	Germany	1900	Erika. Bijou. *See* Chap. 3 (Erika)
Imperator	S	Germany	1924	Urania
Imperial	P	Italian		Oliver

Name	Electric, Standard or Portable	Country of Origin	Date	Other Names if any; Remarks
Imperial	S	Germany	1900	Only a few made
Imperial	S	U.K.	1908	*See* Chap. 4 (Imperial)
Imperial '50'	S	U.K.	1927	*See* Chap. 4 (Imperial)
Imperial Portable	P	U.K.	1930	Torpedo. Mead. Regent. *See* Chap. 4 (Imperial)
Index Visible	P	U.S.A.	1901	
Industria	P	Italy	1923	Juventa
Industrie	S	Germany	1914	Commercial
★International	S	U.S.A.	1889	
Invicta	P	Italy	1938	Olivetti
Iris	P	Italy	1960	Montana
Iskra	S	Poland	1924	Idea
Ivrea	P	Italy	1931	Olivetti Portable, sold in Tel-Aviv
J				
★Jackson	S	U.S.A.	1898	
Janus	P	Germany	1911	Meteor
★Japanese	S	Japan	1919	
★Japy	S	France	1910	Now Hermes
Japy	P	France	1931	Now Hermes
★Jewett	S	U.S.A.	1892	
Junior	P	U.S.A.	1907	
Junior Empire	P	Switzerland	1937	Hermes
Junior-Riter	P	U.S.A.		Portable Remington
Juventa	P	Italy	1922	Later Everest
Juwel	P	Germany	1935	Flott. Dankers
K				
Kamo	S	Germany	1909	Minerva
★Kanzler	S	Germany	1903	
★Kappel	S	Germany	1914	
Kenbar	P	Germany	1935	Rheinmetall
★Keystone	P	U.S.A.	1898	
Klein-Adler	P	Germany	1913	*See* Chap. 3 (Adler)
Kneist	Indicator	Germany	1893	
Knoch	S	Germany	1895	Ford
Koh-I-Noor	S	Germany	1909	Triumph
Kolibri	P	Germany	1955	Groma
Kolombus				Mercedes. *See* Chap. 7
Komet	S	Germany	1913	
★Kosmopolit	Indicator	Germany	1888	

Name	*Electric, Standard or Portable*	*Country of Origin*	*Date*	*Other Names if any; Remarks*
L				
La Standard	Junior	Germany	1948	Passat
*Lambert	Indicator	U.S.A.	1883	
L. C. Smith	S	U.S.A.	1888	*See* Chap. 5 (Smith-Corona)
Leader	P	U.S.A.	1933	Underwood
Leframa	S	Germany	1912	Faktotum
Leggatt	P	U.S.A.	1922	Rochester
Lemco	S	Germany	1912	Minerva
Leningrad	S	Russia	1931	
Lettera 22	P	Italy	1950	Olivetti
Lexicon 80	S	Italy	1953	Olivetti
Libelle	S	Germany	1921	Reliable
*Liberty	Junior	U.S.A.	1922	Molle
Libia	S	Germany	1912	Minerva
Liga	S	Germany	1921	Reliable. Phoenix. Minerva
Lignose	S	Germany	1924	
Lilliput	P	Germany	1907	Also a British toy machine
*Linowriter	S	U.S.A.	1910	Smith Premier
Littoria	S	Italy	1934	Invicta. Juventa
Lloyd	S	U.K.	1912	Imperial
Longini	Signwriter	Belgium	1906	
Lusta	S	Germany	1912	Minerva. Reliable
Lutece				Olympia
Luxa	P	Italy		Sim
M				
Mafra				Commercial. Reliance
*Manhattan	S	U.S.A.	1898	
Manifold				Smith Premier
*Map	S	France	1921	
Mas	P	Italy		Oliver Portable. Sim
*Maskelyne	S	U.K.	1890	
*Masspro	P	U.S.A.	1932	
Master				Royal. Remington. Underwood
Matous	P	Germany	1935	
Matura				Triumph. Electric
McCall	S	U.S.A.	1906	Hooven
*McCool	P	U.S.A.	1903	
Media	S	Germany	1914	Commercial. Hermes
Melbi	Blind	Germany	1919	Gerda
Melior	P	Germany	1911	Meteor
Melitta	P	Germany	1933	Mercedes Portable
*Mentor	Junior	Germany	1909	

Name	*Electric, Standard or Portable*	*Country of Origin*	*Date*	*Other Names if any; Remarks*
Mepas	S	Germany	1914	Commercial
*Mercantile	S	U.S.A.	1908	American. Fleet. Elgin
Mercedes	S	Germany	1907	
Mercedes Elektra	S	Germany	1923	*See* Chap. 7 (Mercedes)
Mercedes Portable	P	Germany	1923	As Underwood
Mercurius	S	Germany	1912	Minerva
Mercury	P	U.K.	1887	
Merkur	P	Germany	1908	Phönix
*Merritt	P	U.S.A.	1890	
*Merz	P	Germany	1926	
Metall	S	Germany	1921	Rheinmetall
Meteko	S	Germany	1912	Minerva
Meteor	P	Germany	1911	
Mignon	Indicator	Germany	1904	Yu Ess. Olympia
Mimeograph	S	U.S.A.	1894	
Minerva	S	Germany	1912	*See* Harris
Minimax	S	Germany	1907	
Miran	P	Italy	1950	Montana
Mitex	Junior	Germany	1922	
Mockba (Moskva)	P	Russia		Portable (Continental?)
*Molle	Junior	U.S.A.	1906	
*Monarch	S	U.S.A.	1904	Smith Premier 4
*Monarch-Pioneer	P	U.S.A.	1920	Remington Portable
Monica	S	Germany	1924	Also name of modern Olympia Portable
Monofix	S	Germany	1921	Mentor
Monopol	S	Germany	1912	Minerva
Montana	P	Italy	1950	Antares
*Moon Hopkins	S	U.S.A.	1902	
*Morris	P	U.S.A.	1887	
Moya	P	U.K.	1902	Early Imperial
Moyer	S	U.S.A.	1916	Blick-Bar. H. Smith
*Munson	Junior	U.S.A.	1890	Chicago. Draper
Musicwriter	S	U.S.A.		
M–W	P	Germany	1924	Gundka. Frolio

N

Name	*Electric, Standard or Portable*	*Country of Origin*	*Date*	*Other Names if any; Remarks*
Naco	S	Germany	1923	Venus
Nagri	P	Germany	1923	Orga-Privat
*National	S	U.S.A.	1889	
*National	P	U.S.A.	1917	
Nauco	S	Germany	1923	Venus
Naumann & Co.	S	Germany	1923	Venus

Name	Electric, Standard or Portable	Country of Origin	Date	Other Names if any; Remarks
*New Century	S	U.S.A.	1900	Caligraph
New England	Indicator	U.K.	1900	
Neya	P	Germany	1925	
Niagara	Indicator	U.S.A.	1902	
*Nickerson	S	U.S.A.	1907	
Nippo	P	Japan	1966	Sold U.K. as The Nippo
*Nippon	S	Japan	1915	Company still produce Portables
Noiseless	S	U.S.A.	1912	Later Remington. Underwood & Smith Premier. Kidder
*Noiseless	P	U.S.A.	1921	
Nora	S	Germany	1914	
Nord	S	U.K.	1892	North
Norden	P	Sweden	1928	Halda
Nordisk	S	Denmark	1918	
Norica	S	Germany	1907	
*North's	S	U.K.	1892	Nord
Nova	P	U.S.A.	1885	Sun
Nuova-Lfui	P	Italy		Oliver. Sim

O

Name	Electric, Standard or Portable	Country of Origin	Date	Other Names if any; Remarks
*Odell	Indicator	U.S.A.	1889	
Odo/Odoma	S	Germany	1921	Blickensderfer-Odoma. Dactyle
Office-Writer				Remington Portable
Official	P	U.S.A.	1901	
Ohio	Junior	U.S.A.	1890	Munson
Oliver	S	U.K.	1928	3-Bank—4-Bank 1935 onwards. *See* Chap. 7
Oliver Portable	P	U.K.	1931	*See* Chap. 7
Oliver Model 20	S	U.K.	1935	Halda-Norden 2. *See* Chap. 7
Oliver 4-Bank	S	Denmark	1929	
Olivetti	S	Italy	1911	*See* Chap. 4
Olivetti Lettera	P	Italy/Spain	1950	*See* Chap. 4
Olivetti Lexicon	S	Italy	1948	And England. *See* Chap. 4
Olivetti Studio 42	Junior	Italy/Spain	1950	*See* Chap. 4
Olivetti Studio 44	Junior	Italy/Spain	1953	*See* Chap. 4
Olympia	S	W/E. Germany	1930	*See* Chap. 5
Olympia	P	W/E. Germany	1931	*See* Chap. 5
Omega	P	Germany	1919	Franconia
Omega	S	Germany	1911	
Optima	S	E. Germany	1950	*See* Chap. 4

Name	Electric, Standard or Portable	Country of Origin	Date	Other Names if any; Remarks
Optima	P	E. Germany	1950	*See* Chap. 4
Orbis	S	Germany	1914	Commercial
Orga	S	Germany	1922	Bing
*Orga-Privat	P	Germany	1923	Bing
Orientans				Culema
Orplid	S	Germany	1914	Commercial
Otto	Vacuum	U.S.A.	1907	
P				
Pacior	P	Poland	1921	
Paeva	P			Antares
Pagina	P	Germany	1911	Meteor
Parisienne	Indicator	France	1886	
Passat	Junior	Germany	1948	
Patria	P	Switzerland	1936	
Patria	P	Spain	1949	Also Japy. Beaucourt & Oliver
Pearl	Indicator	U.S.A.	1893	Peoples
*Peerless	S	U.S.A.	1891	
*Peirce Accounting Machine	S	U.S.A.	1912	
*Peoples	Indicator	U.S.A.	1893	
Perfect	S	U.K.	1892	Salter
Perfekt				Triumph
Perkeo	P	Germany	1912	Urania
Perlita	P	Germany	1924	Gundka. Frolio
Petite				Toy machine
Phönix	P	Germany	1908	Merkur
Phönix	S	Germany	1924	Hega
Piccola				Corona. Erika. Urania
*Picht	P	Germany	1899	For use for the blind
*Pittsburgh	S	U.S.A.	1898	
Plurotyp	Indicator	Germany	1904	Mignon
Plutarch	Indicator	Germany	1904	Mignon
Polyglott				Ideal
Polygraph	P	Germany	1903	
Portex	P	U.S.A.	1922	National
*Porto-Rite		U.S.A.	1935	Remington
*Postal	P	U.S.A.	1903	
Presto	P	Germany	1921	Senta
Princess	P	Germany	1948	
Proctor	S	Germany	1924	
Progress	S	Germany	1921	Also name of Olympia Portable

Name	Electric, Standard or Portable	Country of Origin	Date	Other Names if any; Remarks
Protos	S	Germany	1922	Mercedes, when sold in England
Protos Portable	P	Germany	1924	
Pullmann	S	U.S.A.	1899	American

The Question of 'Q'

It is strange that no typewriter manufacturer has ever thought of a name beginning with 'Q' for a make of machine, particularly as most of the early producers, and many of the more recent ones, seem to have spent a great deal of their time and money squabbling between themselves about names of machines such as Sholes, Remington, Corona, and many others. With injunctions and writs flying in all directions, it does seem rather odd that a machine with a name commencing with the letter 'Q' was never made.

True, the Remington Company had a portable called the 'Quietriter' and there was an old 'Royal' portable by the name of 'Quiet-Deluxe', but these machines were primarily and exclusively advertised and sold under the manufacturers' names of 'Remington' and 'Royal'. These were the only two.

For some peculiar reason 'Q' seems to have been completely ignored. Many good names rapidly spring to mind, certainly much better than some that were used. 'Quadruple' or 'Quartet' could easily have been applied to early four-bank machines. Then others, such as 'Quality', 'Quaver', 'Queen' (after all there is a 'Royal' and an 'Imperial'). 'Quick', Queensbury', 'Quest'. 'Quiet'—such an obvious one! 'Quill' is another possible, and perhaps 'Quiver' with an arrow through the centre, indicating speed. In a lighter vein, many of them would have been well advised to call their machines 'Quagmire', in which most of them seem to have been from the very beginning! or perhaps 'Quixote', as most of them were tilting at windmills anyway. Some were certainly 'Queer'.

It is of no importance, but the fact remains that a 'Q' has never been used to start the name of a Typewriter Manufacturing Company, whereas every other letter has. There was even an 'Xcel' and a machine called the 'Zerograph'. Ironically, the first letter on the existing inefficient keyboard known as 'QWERTY' is 'Q'.

R

Name	Electric, Standard or Portable	Country of Origin	Date	Other Names if any; Remarks
Radio	S	Germany	1909	Dea
Rally 220				Rooy
*Rapid	P	U.S.A.	1888	Juwel
R. C. Allen				Woodstock
Regent	S	Germany	1909	Minerva (Imperial in England)
Regent	S	Germany	1907	Torpedo when sold in England (also Imperial)
Regent	P	U.K.	1930	Imperial (assembled U.K.)

Name	*Electric, Standard or Portable*	*Country of Origin*	*Date*	*Other Names if any; Remarks*
Regina	S	Germany	1904	
Reliable	S	Germany	1921	
*Reliance-Premier	S	U.S.A.	1917	Pittsburgh Visible
Rembrandt	S	Holland		Remington Masteriter
Remette	P	U.S.A.	1938	Remington
Remington	S	U.S.A.	1873	*See* Chap. 1 (Glidden Sholes) and Chap. 5 (Remington)
Remington Electric	S	U.S.A.	1925	do.
Remington Noiseless	S	U.S.A.	1924	do.
Remington Portable	P	U.S.A.	1923	do.
Remington-Sholes	S	U.S.A.	1896	do.
Remington-Sholes	S	U.S.A.	1902	do.
Remington Super	P	U.S.A.	1950	do.
*Reporters Special	S	U.S.A.	1911	*See* Harris
Revolution				Minerva
Rex	S	U.S.A.	1911	Autocrat. Harris Reporter
Rheinita	S	Germany	1929	
*Rheinmetall	S	Germany	1920	
Rival	S	Germany	1912	Minerva
*Roberts Ninety	P	U.S.A.	1919	Blick Ninety
Rochester	P	U.S.A.	1923	
Rocket	P	Switzerland	1951	Hermes Baby
*Rofa	S	Germany	1921	
Roland	S	Germany	1912	Minerva
Rooy	S	France	1937	
Rooy Portable	P	France	1950	
Rooy Standard	S	France	1946	
Royal	S	U.S.A.	1906	*See* Chap. 5 (Royal)
Royal Barlock	S	U.K.		*See* Chap. 7 (Barlock)
Royal Electric	S	U.S.A.	1950	*See* Chap. 5 (Royal)
Royal Express	S	U.K.		Salter
Royal Portable	P	U.S.A.	1926	*See* Chap. 5 (Royal)
Roxy	S	France	1954	

S

Name	*Electric, Standard or Portable*	*Country of Origin*	*Date*	*Other Names if any; Remarks*
Sabaida	S	Italy	1921	Invicta
Sabb	P	Germany	1928	Juventa. Later Everest (*See* Chap. 7)
Saleem				Urania
*Salter	S	U.K.	1892	
*Salter	S	U.K.	1907	
Sampo	P	Sweden	1894	
Sarafon				Minerva

Name	*Electric, Standard or Portable*	*Country of Origin*	*Date*	*Other Names if any; Remarks*
⋆Saturn	S	Switzerland	1897	
Saxonia	S	Germany	1921	Reliable
Schade	Sphere	Germany	1896	
Schapiro	Indicator	Germany	1894	
Scout	P			Remington
Scribe	P	U.K.		Olivetti
Scripta	P		1924	Gundka. Frolio
Scriva	P	Germany	1928	Triumph when sold in Sweden
⋆Secor	S	U.S.A.	1905	
Selectric	E	U.S.A.		IBM '72'
Senator	S	Germany	1922	
⋆Senta	P	Germany	1914	Presto-Balkan
Shilling	S	U.S.A.	1921	
Shimad		Japan	1938	
⋆Sholes-Visible	S	U.S.A.	1901	Zalmon G. 1911
⋆Sholes Zalmon	S	U.S.A.	1911	Zalsho. Acme. Waterbury
Shortwriter	P	U.S.A.	1914	Shorthand writing machine
Siegfried	S	Germany	1913	
Siemag	S	Germany	1949	Later Messa, Portugal. *See* Chap. 4
Siemag Portable	P	Portugal/ Germany	1954	Later Sterling, U.K.
Signet	P	Japan		Royal. Imperial
Silent	S	U.S.A.	1936	Corona
Sim	P	Italy	1931	Juventa. Everest
Simplex		U.S.A.	1901	Toy machine
Simplex	S	U.S.A.	1909	
Skandia				Norden
Skyriter	P	U.S.A.		Smith-Corona
Smith Corona	P	U.S.A.	1931	*See* Chap. 5
Smith Corona Electric	S	U.S.A.	1955	*See* Chap. 5
⋆Smith (Emerson)	S	U.S.A.	1907	
Smith L. C.	S	U.S.A.	1904	*See* Chap. 5
Smith Premier	S	U.S.A.	1889	*See* Chap. 5
Smith Premier Noiseless	S	U.S.A.	1925	Remington. Underwood
⋆Smith Visible	S	U.S.A.	1909	
Speedking	P	U.S.A.	1938	Royal
Speedster	S	U.K.	1922	Oliver Model 11
Sphinx	S	Switzerland	1913	
Stallman	Indicator	Germany	1918	Yu Ess. Mignon
Standard	P	U.S.A.	1938	Corona
Standard Folding	P	U.S.A.	1907	Later 3-bank Corona
Star			1905	Sun. Halda
⋆Stearns	S	U.S.A.	1905	

Name	Electric, Standard or Portable	Country of Origin	Date	Other Names if any; Remarks
Stella	Indicator	Germany	1918	Yu Ess. Mignon
*Stenograph	P	U.S.A.	1889	Shorthand machine
Stenotype	P	U.S.A.	1911	Shorthand machine
Sterling	P	U.S.A.	1932	Corona
*Sterling	P	U.S.A.	1910	Production ended 1913
Sterling Siemag		Portugal		Mesa
Stoewer	S	Germany	1903	Swift. Conqueror
Stoewer Elite	P	Germany	1912	
Stoewer Portable	P	Germany	1926	
Stolzenberg	S	U.K.	1906	Oliver
Streamliner	P	U.S.A.		Remington
Studio	P	Italy	1938	Olivetti
*Sun	P	U.S.A.	1885	Star. Nova. Carlem. Lefrala.
Super				Rheinmetall
Superba	P	Germany	1936	Mercedes
Swift	S	Germany	1904	Stoewer when sold in England
Swift-Record	S	Germany	1904	do.
Swissa	P	Switzerland	1950	
T				
*Taurus	P		1908	Only a few made
Tell	P	Germany	1922	Mitex
Tempotype	S	Germany	1909	Triumph when sold in Denmark
Thuringen				Commercial. Minerva
Thuringia				Mentor. Reliable
Tip-Tip		Czechoslovakia	1936	Mignon
Tippa (Gossen)	P	Germany	1948	Later Adler Tippa
*Titania	S	Germany	1910	
Tops		Yugoslavia		
Torpedo	S	Germany	1910	Bluebird (England)
Torpedo Portable	P	Germany	1949	Bluebird (England)
Transatlantic	S	Germany	1923	Venus
*Travis		U.S.A.	1905	
Triumph	S	Germany	1909	*See* Chap. 5
Triumph Portable	P	Germany	1929	*See* Chap. 5
Typo		U.K.	1919	Imperial Model 'D'
Typorium		Germany	1904	A.E.G. Mignon
U				
Uhlig	S	U.S.A.	1910	Associated (Woodstock)
Ujlaki	S	Germany	1935	Urania

Name	Electric, Standard or Portable	Country of Origin	Date	Other Names if any; Remarks
Ukrainia	S	Russia	1935	Made in Kiev
Ultima	S	Germany	1909	Helios
Unda	P	Austria	1923	
Underwood	S	U.S.A.	1896	*See* Chap. 7
Underwood Electric	S	U.S.A.	1947	*See* Chap. 7
Underwood Noiseless	S	U.S.A.	1930	Remington Noiseless
Underwood Seg/Shift	S	U.S.A.	1946	
Union	S	Germany	1908	Dea
Unitype	S	Germany	1907	Torpedo
Urania	S	Germany	1910	
Urania-Piccola	P	Germany	1935	
Ur Du	S	Germany	1910	Urania
Us Mirsa-Ideal	S	Germany	1910	Ideal
V				
Varsity	P	U.S.A.	1939	Royal
Vasanta	P	Germany	1923	Meteor
*Velograph	Indicator	Switzerland	1886	
Venus	S	Germany	1923	
Victor	P	U.S.A.	1927	
*Victor I	Indicator	U.S.A.	1889	
*Victor A	S	U.S.A.	1894	
*Victor B	S	U.S.A.	1908	
*Victor 10½	S	U.S.A.	1919	
Victoria	S	Spain	1913	
Vira		Hungary		
Virotyp	Indicator	France	1914	
*Visigraph	S	U.S.A.	1910	
Vittoria	S	Italy	1923	
Volcan	P	Italy		1939 onwards
Volks	Indicator	Germany	1898	
Voss	P	Germany	1949	Later Oliver in U.K.
W				
Wallstreet	S	U.S.A.		Daugherty. Pittsburg
Wanamaker	S	Canada	1912	Kidder principle
Wanderer	S	Germany		Continental
Waterloo	S	Germany	1912	Minerva
*Waverley	S	U.K.	1889	
Wellington	S	U.S.A.	1892	Kidder machine
Weltblick				Blickensderfer
Westphalia	Indicator	Germany	1884	*See* Chap. 7
Wiedmer	P	Germany	1908	

Name	*Electric, Standard or Portable*	*Country of Origin*	*Date*	*Other Names if any; Remarks*
⋆Williams	S	U.S.A.	1892	Two kinds, 3-bank and 4-bank
Wilson	P	Germany	1911	Meteor
Woodstock	S	U.S.A.	1914	Earlier Emerson, later R. C. Allen. *See* Chap. 7
Woodstock Electrite	S	U.S.A.	1924	Similar to Mercedes Electric
Woodstock Seg/Shift	S	U.S.A.	1948	Last Woodstock, later R. C. Allen
⋆World	Indicator	U.S.A.	1886	Blake and Boston
World-Blick				Blickensderfer. *See* Chap. 7
⋆Wright Speedwriter	S			No known source
Write-Easy	P	Germany	1924	Frolio, Gundka, etc.
X				
⋆Xcel	S	U.S.A.	1922	
Y				
Yanalif	S	Turkey	1929	Agamli-Ogli
⋆Yetman	S	U.S.A.	1896	Transmitting
Yost		U.S.A.	1887	*See* Chap. 7
Yu Ess	Indicator		1918	Mignon. Olympia. A.E.G.
Z				
Zalsho	S	U.K.	1913	
Zalsho	S	U.S.A.	1915	
Zephyr	P	U.S.A.	1938	Corona
Zeta	S	Czechoslovakia	1949	Consul Diplomat
Zeta Portable	P	Czechoslovakia	1951	Consul. *See* Chap. 5
⋆Zerograph	S, E	U.K.	1895	Forerunner of Telex

⋆ Machines marked with an asterisk will also be found in the following illustrations (Figures 8.13–8.179) between pages 246 and 265.

J

8.13

8.14

8.15

8.13 ALEXANDER 'A' (1914)
Alexander Typewriter Company, U.S.A.

8.14 ALEXANDER 'B' (1923)
Company unknown, Los Angeles, U.S.A. (Group 10)

8.15 ALLEN 3-BANK (1919)
Allen Typewriter Company, U.S.A. Three-bank (Group 10)

8.16

8.18

8.17

8.16 ALLEN 4-BANK (1920)
Allen Typewriter Company, U.S.A. (Group 10)

8.17 AMERICAN 'B' (1901)
(Group 1)

8.18 AMERICAN POCKET (1926)
Chicago, U.S.A. Few only produced

8.19

8.20

8.21

8.19 AMERICAN STANDARD (1892)
Remington contested the name STANDARD and it was then called JEWETT, U.S.A. (Group 6) Up-strike

8.20 AMERICAN VISIBLE (1893)
American Typewriter Company, New York, U.S.A. Had rubber type—very few produced

8.21 ARMSTRONG (British) (1907)
British Typewriter Company, Abington, England. Identical with American

8.22

8.23

8.24

ARNOLD (1876)
Patent model only
AUTOCRAT (1916)
Rex Typewriter Company, Wisconsin, U.S.A. (Group 10). Three-bank
AUTOMATIC (1883)
Major E. M. Hamilton, U.S.A. (Group 6). Up-strike. Finished 1887. Also known as HAMILTON

8.25

8.26

8.27

BARR (1926)
U.S.A., Canada. (Group 10)
BENNETT (1910)
U.S.A. (Group 2). Known as JUNIOR ELLIOTT-FISHER
BIJOU 'A' (1910)
Seidel Naumann, Dresden, Germany. Ideal. (Group 9). Three-bank model

8.28

8.29

8.30

BING (German) (1925)
Bingwerke Nuremberg, Germany. No. 1 employed ink pad
BING 2 (1927)
Germany. Also known as AMKA, BERNI, and BEKE. (Group 9). Used a ribbon
BOSTON (1888)
Invented by John Becker, Patent 1886. U.S.A. (Group 5)

8.31

8.32

8.31 BRADY WARNER (1878)
U.S.A. No other details. (Group 6) Down-strike from rear

8.32 BROOKS (1885)
United Typewriter and Supplies Company, New York, U.S.A.

8.33 BURNETT (1908)
U.S.A. Also known as IMPERIAL VISIBLE and TRIUMPH VISIBLE. (Group 6)
Down-strike from rear

8.34

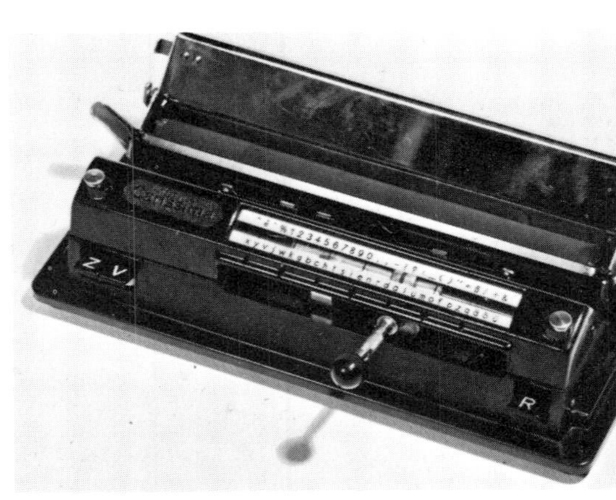

8.35

8.34 BURNS (1889)
Burns Typewriter Company, U.S.A. (Group 6) Up-strike. Very few machines manufactured, but Company continued making type for other more successful typewriters

8.35 CALIGRAPH 1 (1881)
U.S.A. (Group 6) Up-strike

8.36 CARISSIMA (1934)
Germany. (Group 2). Very small, would fit in pocket. Few only were made

8.37

8.38

8.3

8.37 CASH (1887)
U.S.A. (Group 6) Down-strike from front. Also known as the TYPOGRAPH

8.38 CENTURY (1919)
American Manufacturing Company, U.S.A. Known as the REMINGTON JUNIOR. (Group 10). Three-bank

8.39 CHAMPION
Garvin Machine Company, U.S.A. Also known as PEOPLES-CHAMPION and PEARL. (Group 2) Type-Wheel

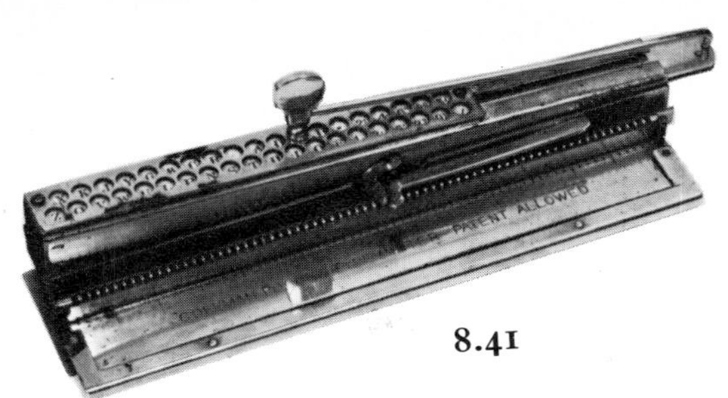

8.41

8.42

8.40

o CHICAGO (1898)
Chicago Writing Machine Company, U.S.A. Also known as CONOVER, DRAPER, BALTIMORE, GALESBURG. (Group 2)

1 COFFMAN (1902)
U.S.A. Type on rubber belt. Few were made

2 COLUMBIA (1900)
Columbia Typewriter Company, U.S.A. (Group 2) Type-wheel

8.43

8.44

8.45

43 COMMERCIAL VISIBLE (1898)
U.S.A. Also called POSTAL. (Group 2) Type-wheel. Three-bank. Very few made

44 LUCIEN CRANDALL'S WRITING MACHINE (1881)
U.S.A. (Group 2). Had neither typebars nor wheel but fixed type in half a dozen bands around a cylinder. When the operator depressed a key, the cylinder shifted laterally and rotated selecting the proper band and the desired letter

45 CRANDALL 3 (1906)
U.S.A. (Group 3) Type-sleeve

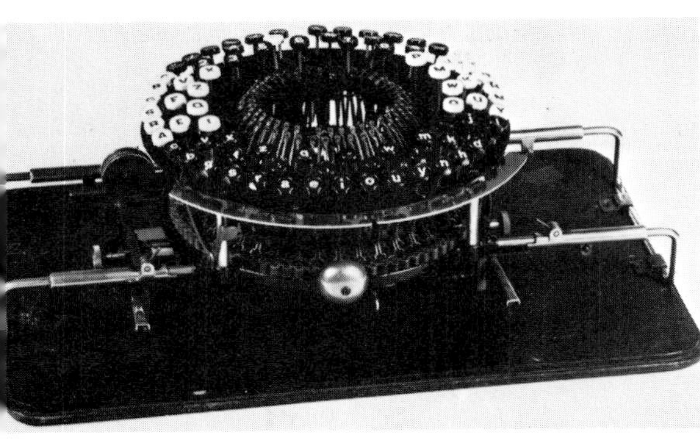

8.46

8.47A

8.47B

.46 CRARY (1892)
U.S.A. (Group 4/6) Circular Down-strike

.47A CROWN (1894)
National Meter Co., U.S.A. (Group 2)

.47B CROWN PORTABLE (1911)
Chicago, U.S.A. (Group 9) Three-bank

8.48

8.50

8.49

8.48 DAUGHERTY (1891)
U.S.A. Also sold as PITTSBURGH-RELIANCE, RELIANCE-PREMIER, FORT PITT, and SHILLING. (Group 10) Front-strike, with four rows of keys, and was the forerunner of all modern front-strike typewriters

8.49 DAW AND TATE (TAIT) (1884)
England. A circular down-strike machine. Only two were ever made

8.50 DAYTON (1924)
Dayton Portable Typewriter Company, U.S.A. (Group 10). Four-bank. Very few were manufactured

8.51

8.52

8.53

8.51 DECKER-BEACHLER (1917)
U.S.A. Other names are PITTSBURGH-RELIANCE, RELIANCE-PREMIER, and AMERICAN. (Group 10) Front-strike. Four rows of keys, and was the successor of the PITTSBURGH. Production ceased in 1921

8.52 DEFI (1908) Eagle Typewriter Company, U.S.A. Also sold as the EAGLE. (Group 1) Swinging Sector

8.53 DEMOUNTABLE (1921) Demountable Typewriter Company, U.S.A. (Group 10) Front-strike Design. Four-bank keyboard. Ceased production in 1936

8.54

8.55

8.56

8.54 DENNIS DUPLEX (1895) Duplex Typewriter Company, U.S.A. (Group 6) Up-strike. Ten rows of keys. In this machine an attempt was made to write combinations of words by striking two keys at the same time. This proved unsuccessful

8.55 DENSMORE (1891) U.S.A. (Group 6) Up-strike. A slight shifting movement brought the writing into view. It was the first typewriter to use the platen release, which was later adopted on all typewriters

8.56 DENSMORE 6 (1907) Union Typewriter Company, U.S.A. (Group 6) Up-Strike. A tabulator was added

8.57

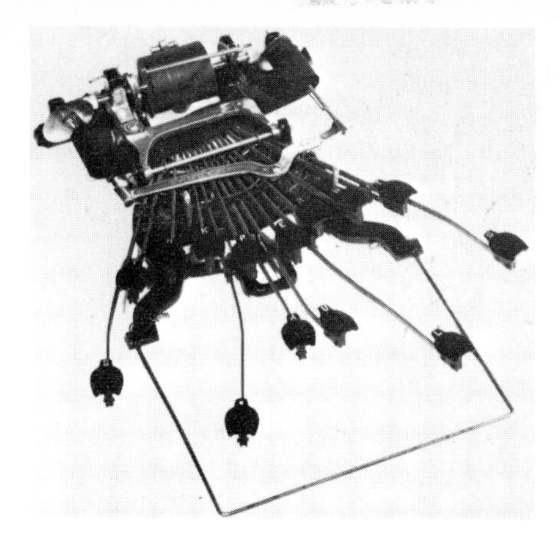

8.58

5.59

7 DIAMOND (1923)
Fred Sholes. (Group 10). Was only an experimental machine

3 DICTATYPE (1888)
Dictatype Shorthand Machine Company, U.S.A. No classification. Only a few were ever constructed

) D.M.G.
Manufactured by Mercedes-Benz Aktiengesellschaft under the name of DAIMLER-MOTOREN-GESELLSCHAFT, thus D.M.G. It is unknown how many of these machines were manufactured. Production commenced early 1920s and was discontinued 14 November 1925. Only one known model exists and this is in the Daimler-Benz Motor Museum in Stuttgart. The patents to manufacture the machine were issued after production had ceased. Some authorities state the machine tools were disposed of to a Russian source, though the Mercedes-Benz Museum have no record of this. The carriage was detachable and interchangeable, and the machine was of modern design and construction. Photo-copies of the patents can be seen at the British Typewriter Museum, Bournemouth. We are deeply indebted to the Mercedes-Benz Manufacturing Company for this information. This machine was in no way connected with the MERCEDES typewriter

•

8.61

8.62

o ECKELS (1892)
Chicago, U.S.A. Seven rows of keys. Very little else known about this machine

1 ECLIPSE (about 1885)
Similar to the BROOKS. Manufacturers unknown

2 EDELMANN (1897)
Germany. (Group 2) Type-Wheel Design

8.64

8.65

8.64 ELLIOTT-FISHER (1903)
U.S.A. A development of the Book-writing principle

8.65 ELLIOTT-HATCH (1897)
U.S.A. (Group 6) Down-strike. This was a Book-writing Machine designed specifically for writing on flat surfac

8.63

8.66

8.67

8.63 EDISON-MIMEOGRAPH (1894)
Thomas A. Edison. (Group 2) Type-Wheel Design. It was not a commercial success. It survived for only a brief period

8.66 ENGLISH (1890)
British manufacture. Two rows of keys. (Group 6) Down-strike from front

8.67 FAY-SHOLES (1901)
Remington-Sholes/Fay-Sholes. U.S.A. (Group 6) Up-strike

8.68

8.69

8.70

8.68 FITCH (1886)
Brady Manufacturing Company, U.S.A. (Group 6) Down-strike from rear

8.69 FORD (1895)
Ford Typewriter Company, U.S.A. Also known as the KNOCH and HURTU. Few only sold. (Group 4) Radial-strike

8.70 FOX (1906)
Fox Typewriter Company, U.S.A. Also sold as the RAPID and the FOX VISIBLE. (Group 9) Front-strike Design with staggered bearings. Went into receivership in 1921

8.71

8.72

8.73

71 FOX (BABY PORTABLE) (1918)
Fox Machine Company, U.S.A. (Group 10) Front-strike. Three-bank keyboard. Ceased production after litigation with Corona Typewriter Company

72 FOX BLIND (1898)
Fox Typewriter Company, U.S.A. (Group 6) Up-strike. Ceased production 1914

73 FOX STERLING (1921)
Fox Typewriter Company, U.S.A. Three-bank, double shift. Few were produced

8.74

8.75

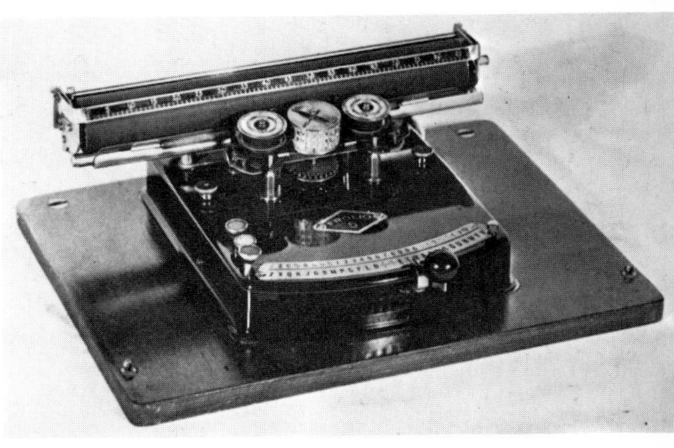

8.76

·74 FRANKLIN (1889)
Franklin Typewriter Company, U.S.A. (Group 6) Down-Strike from Front

·75 FRISTER and ROSSMAN (Copy of Caligraph, U.S.A.) (1892)
Frister and Rossmann, Berlin, Germany. (Group 6) Under-strike

.76 FROLIO (1924)
Gundka Werke, Brandenburg, Germany. Known as G. and K. GUNDKA, M-W, PERLITA, WRITE EASY. (Group 2) Type-Wheel/(Group 5) Index System Designs

8.77

8.78

8.79

3.77 GARBELL (1919)
Garbell Typewriter Company, U.S.A. (Group 4). Three-bank. Bankrupt 1923

3.78 GARDNER (1893)
Gardner British Typewriter Company, Manchester, England. (Group 3). The letters were on small squares of india-rubber. As it only had sixteen key tops, two keys had to be depressed or a combination of keys depressed to print the letter required as with a musical instrument. It is stated that 'the touch of the GARDNER was something fearful to contemplate'. It soon found its way into the hands of the liquidator in 1895

3.79 GENIATUS (1928)
Manufactured in Germany. A copy of the AMERICAN. (Group 1). Manufactured principally for mail-order tra

8.80

8.81

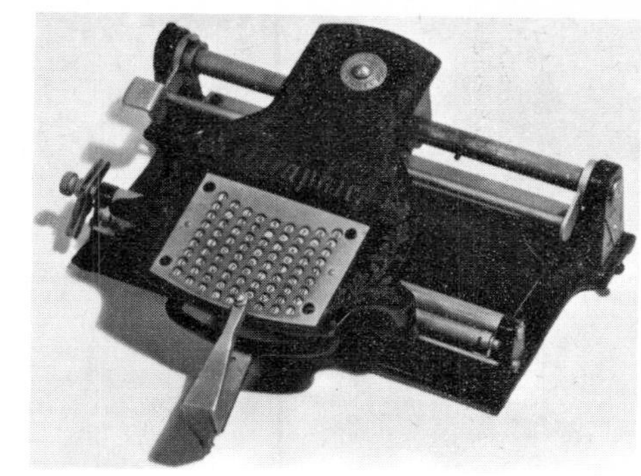

8.82

8.80 GOURLAND (1920)
Gourland Typewriter Corporation, U.S.A. (Group 10) Front-strike. Four-bank. 1925 production was discontinued. It was said production was to be transferred to Russia. Nothing further was known

8.81 GRANVILLE AUTOMATIC (1896)
Granville Manufacturing Company, U.S.A. (Group 4) Forward-strike. Four rows of keys. Production later transferred to England with little success

8.82 GRAPHIC (1895)
C. F. Kindermann and Company, Berlin, Germany. Similar to the HALL

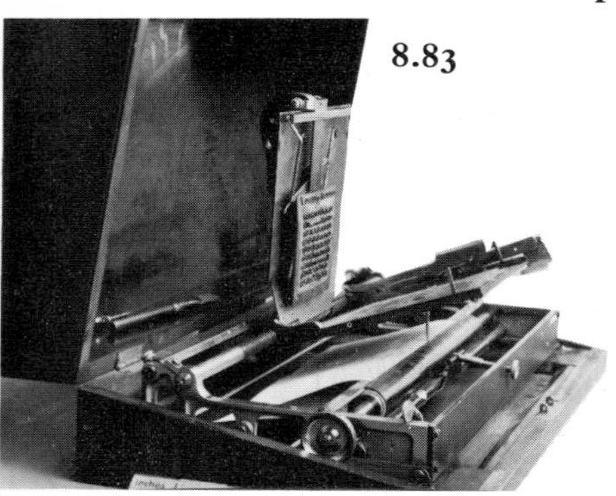

8.83

8.84

8.85

8.83 HALL (1881)
Thomas Hall, Brooklyn, U.S.A. (No Grouping). Rectangular metal guide plate. Each letter required two operations. Only about 1,000 of these were manufactured

8.84 HALL BRAILLE WRITER (1891)
Munson Typewriter Company, U.S.A. Invented by Frank H. Hall. Embossed the alphabet on paper

8.85 HAMMONIA (1883)
Guhl and Harbeck of Hamburg, Germany. (Group 5) Index System. Very slow in

8.88

8.87

8.86 HANSON (1899)
Nickerson Typewriter Company, Chicago, U.S.A. Radically different. Just an impractical experimental machine. Never produced

8.87 HARRIS (1911)
Harris Typewriter Company, U.S.A. (Group 10) Front-strike. Three-bank, double shift. Originally made for Sears Roebuck Mail Order Company. Also sold as the REX, and later became associated with the DEMOUNTABLE

8.88 HARRIS VISIBLE 4 (1914)
Rex Typewriter Company, U.S.A. (Group 10) Front-strike. Three-bank. Almost identical with the HARRIS

8.89

8.90

8.91

89 HARTFORD (1894)
John M. Fairfield, Hartford, U.S.A. Also known as the CLEVELAND. (Group 6) Up-strike. Six rows of keys

90 HARTFORD 3 (1905)
John M. Fairfield, Hartfield, U.S.A. (Group 6) Up-strike. Four-bank keyboard with single shift

91 HOOVEN AUTOMATIC (1912)
Hooven Automatic Typewriter Company, U.S.A. This machine has no grouping.
A perforating machine cut the perforations in the paper roll which was attached to a normal typewriter, and under motor power, repeat operations were run off automatically

8.92

8.93

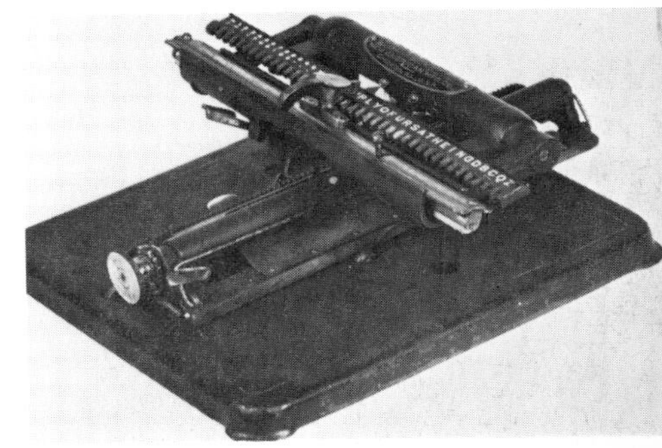

8.94

.92 HORTON (probably about 1883)
This may be an experimental model by Horton, and is different from the model made by E. E. Horton

.93 INTERNATIONAL 'A' (1893)
Manufactured by L. S. Crandall. (Group 6) Up-strike. Four rows of keys

.94 INTERNATIONAL 'B' (1890)
New American Manufacturing Company, U.S.A. A nondescript experimental machine. Inventor—Lucien S. Crandall. Similar to ODELL

8.95

8.96

8.97

8.95 INTERNATIONAL 'C' (1893)
(Group 6) Up-strike. Six-row double keyboard. *The Phonographic World* of New York stated 'Mr. Lucien S. Crandall whose two typewriters the INTERNATIONAL and the CRANDALL, have led to the invention by stenographers of more swear words than all the other typewriters combined'

8.96 JACKSON (1898)
Jackson Typewriter Company, U.S.A. (Group 7) Grasshopper Movement. Similar to the WILLIAMS. Type rested on an ink pad. When the key was depressed it shot forward to the platen. This method instead of using a ribbon was claimed a special feature. It was very short-lived. Had four rows of keys

8.97 JAPANESE (1938)
Shimada Typewriter Works, Tokyo, Japan. It was an index indicator type machine with 2,740 characters on a belt

8.98

8.99

8.100

8.98 JAPY (French) (1910)
Japy Frères and Company, Beaucourt, France. (Group 10) Front-strike

8.99 JEWETT (1892)
Jewett Typewriter Company, U.S.A. (Group 6) Up-strike. Seven-bank keyboard. Discontinued 1910

8.100 JEWETT 1 DUPLEX (1892)
Duplex Typewriter Company, U.S.A. (Group 6) Up-strike. First machine to write two letters at the same time. Seven-bank keyboard

8.101

8.102

8.103

8.101 JEWETT 2 (1895)
Jewett Typewriter Company, U.S.A. (Group 6) Up-strike. Six rows of keys

8.102 KANZLER (1903)
German Typewriter Company. Also known as HANSA, RAPID, and CHANCELLOR. (Group 4) Radial-strike. Obviously Kidder had some influence on this machine

8.103 KAPPEL (1914)
Maschinenfabrik Kappel AG., Chemnitz-Kappel, Germany, from which city the machine derives its name. (Group 10) Front-strike Design. Modern four-bank keyboard

8.104

8.105

8.1

8.104 KEYSTONE (1898)
Keystone Typewriter Company, U.S.A. (Group 1) Swinging Sector Design. Twenty-eight keys, two shift keys, eighty-four characters

8.105 KOSMOPOLIT (German) (1888)
Guhl and Harbeck, Hamburg, Germany. (Group 1) Swinging Sector. Type was cast in two rows on rubber

8.106 LAMBERT (1883)
Gramophone-Typewriter Ltd., & Sister Companies, England. German Gramophone Company, Germany. Also in Dieppe, France. (Group 2). It was an American invention of Frank Lambert of Brooklyn, New York, U.S.A. Lambert is said to have taken seventeen years to perfect his ingenious machine which for a time sold extremely well. Several typists are said to have averaged 110 words a minute—although this is very doubtful

8.107

8.108

8.109

107 LIBERTY (1923)
Liberty Typewriter Company, U.S.A. (Group 10) Front-strike. Three-bank.
Identical with the MOLLE which had failed. The LIBERTY was also unsuccessful

108 LINOWRITER (1910)
Smith Premier, U.S.A. (Group 6) Up-strike. This was a machine with seven rows of keys specially converted to conform to the keyboard of the LINOTYPE machine

109 MANHATTAN (1898)
Manhattan Typewriter Company, U.S.A. (Group 6) Up-strike. Four-bank

8.110

8.111

8.112

.110 MAP (1921)
Manufacture d'Armes de Paris, St. Denis, France. (Group 10) Front-strike. Four rows of keys. The machine had an interchangeable basket

.111 MASKELYNE (1890)
Invention of Messrs. J. N. and Neville Maskelyne of Egyptian Hall fame. It operated with 10 or 12 pitch type, but it was unsuccessful. Very few were made and even less were sold. (Group 7) GRASSHOPPER

.112 MASSPRO (1932)
Mass Production Corporation, U.S.A. (Group 10) Front-strike. Three-bank. Very short lived

8.113

8.114

8.115

.113 McCOOL (1904)
Acme Keystone Manufacturing Company, U.S.A. (Group 2) Type-Wheel. Few were made

.114 MENTOR (1909)
Metallindustrie, Germany. Also known as the MONOFIX and THURINGIA. (Group 10) Front-Strike. Four-bank keyboard

.115 MERCANTILE (1908)
A development of the AMERICAN. Also sold under the names of EAGLE, FLEET, ELGIN, and PULLMAN

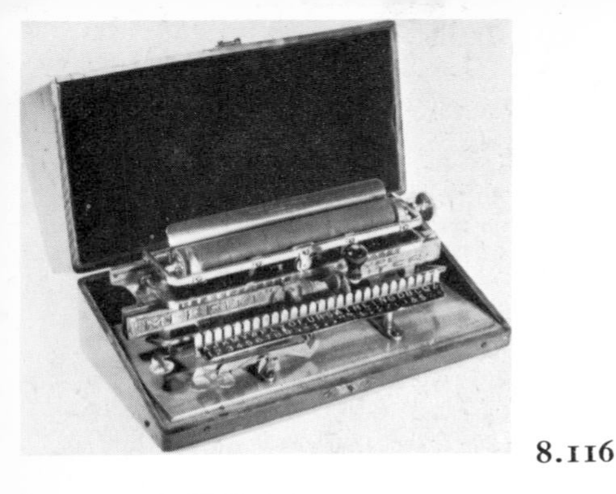

8.116

8.117

8.118

8.116 MERRITT (1890)
Merritt Manufacturing Company, U.S.A. (Group 5) Index System

8.117 MERZ (1926)
Merz-Werke, Frankfurt AM-Main, Germany. Sold also as CONCORD. (Group 10) Front-strike. Four-bank keyboard

8.118 MOLLE (1906)
Molle Typewriter Company, U.S.A. (Group 10) Front-strike. Three-bank

8.119

8.120

8.121

8.119 MONARCH (1904)
U.S.A. (Group 9) Front-strike. Bearings side by side. Four-bank keyboard

8.120 MONARCH PIONEER (1920)
Remington Rand Company, U.S.A. (Group 10) Front-strike. Three-bank

8.121 MOON HOPKINS (1902)
American Arithometer Company, U.S.A. (Group 6) Up-strike. Six rows of keys. It was combined with an adding and listing attachment

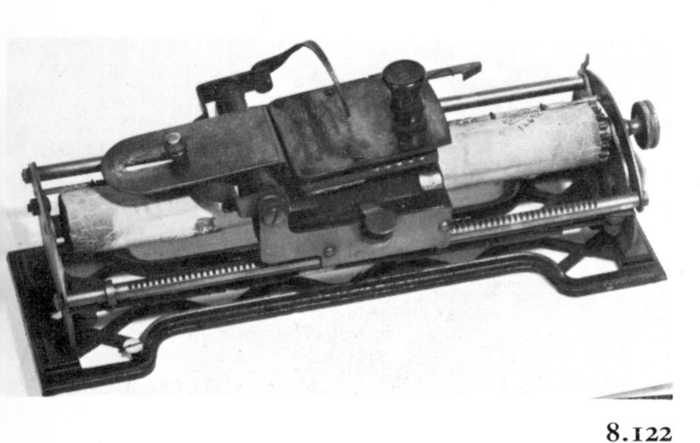

8.122

8.123

8.12

8.122 MORRIS (1887)
Hoggson Manufacturing Company, U.S.A. Similar to the HALL in operation

8.123 MUNSON (1890)
Munson Company, U.S.A. (Group 3). Three-bank keyboard. See CHICAGO

8.124 NATIONAL (1889)
National Typewriter Company, U.S.A. Had a large sale in England and Germany. (Group 6) Up-strike. Invisible writing. There are three rows of keys in a semi-circle

8.125

8.126

8.127

25 NATIONAL 2 (1918)
Rex Typewriter Corporation, U.S.A. (Group 10). Three rows of keys

26 NEW CENTURY CALIGRAPH (1900)
American Writing Machine Company, U.S.A. (Group 6) Up-strike. Seven rows of keys

27 NICKERSON (1907)
Nickerson Typewriter Company, U.S.A. It was unique in that it wrote the short way around the platen instead of across the platen, which was the most radical departure from normal construction. It had four rows of keys. The inventor was Walker Hansen who died at the age of 21. The Rev. Charles S. Nickerson was given a grant by the American Telegraph and Telephone Company. Only one machine was ever produced

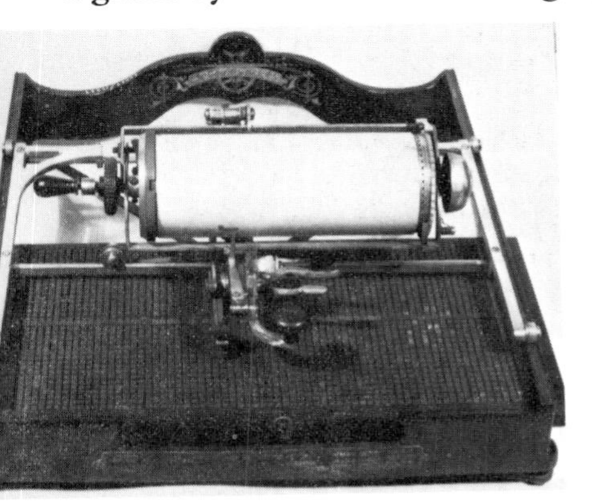

8.128

8.129

8.130

128 NIPPON JAPANESE (1915)
Nippon Typewriter Company Ltd., Tokyo, Japan. The Japanese language is composed of approximately 12,000 characters (although recently the Japanese Ministry of Education has limited the characters to 1,850). These are supplied by a central type case and two auxiliary type cases. The operation is laborious. The type required is picked up by lever and transferred to the paper and returned in the same manner. To the best of our knowledge very few were made

129 NOISELESS PORTABLE (1921)
Produced by the Noiseless Typewriter Company, U.S.A. It was later produced by Remington. (Group 4). Three rows of keys and a double shift. Very few appear to have been sold

.130 NORTH'S (1892)
North's Typewriter Manufacturing Company Ltd., London, England. (Group 6) Down-Strike from Rear. It had four rows of keys and in reality was the older ENGLISH with improvements. The machine would perhaps have enjoyed a longer and much greater success were it not for Lord North's sudden death

8.131

8.132

8.131 ODELL (1889)
Odell Typewriter Company, U.S.A. Made at a very low price Very few were manufactured. Later attempts were made in 1918 when it was revived as th NEW AMERICA without succes

8.132 ORGA-PRIVAT (1923)
Bing-Werke, Nuremberg, Germany. (Group 10). Four-bank

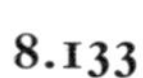

8.133

8.135

8.133 PEERLESS (1891)
Peerless Typewriter Company, U.S.A. (Group 6) Up-strike. Invisible writing with eight rows of keys. It lasted only a few years

8.134 PEIRCE ACCOUNTING (1912)
Peirce Accounting Machine Company, U.S.A. (Group 10). Three rows of keys

8.135 PEOPLES (1893)
Garvin Machine Company, U.S.A. (Group 5) with a type-wheel and a ribbon. Also marketed as the CHAMPION and the PEARL

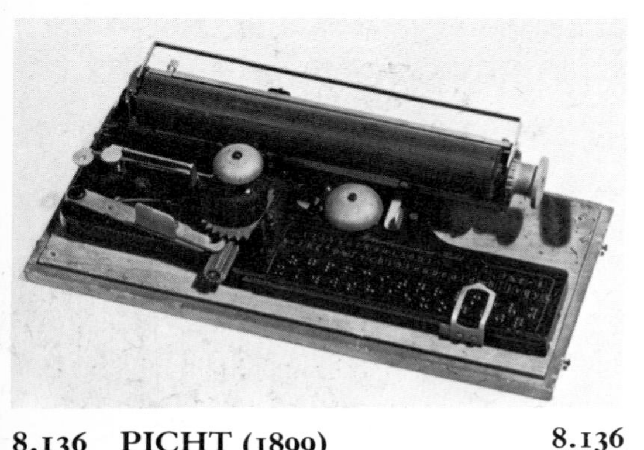

8.136

8.137

8.136 PICHT (1899)
Herde Wendt, Berlin, Germany. (Group 2). A machine for use by the blind

8.137 PITTSBURGH VISIBLE (early 1900s)
Pittsburgh Writing Machine Company, U.S.A. It was an improvement of the DAUGHERTY, and had an interchangeable carriage and type-basket. It was also sold as the RELIANCE-PREMIER, RELIANCE, and the SHILLING. There were four rows of keys. The Company went into receivership in 1913 and production ceased in 1921

8.138 PORTO-RITE (1935)
(Group 10). Four-bank keyboard. It is believed to have been made by the Remington Typewriter Company for a mail-order house. This has not been substantiated

8.139

8.140

8.139 POSTAL (1903)
Postal Typewriter Company, New York, U.S.A. (Group 2) Type-Wheel Design. Three rows of keys

8.140 RAPID (1888)
Manufactured by A. W. Gump & Company. (Group 9). Four rows of keys. An earlier machine used Kidder's forward thrust principle

8.141 RELIANCE-PREMIER (1917)
Pittsburgh Writing Machine Company, U.S.A. (Group 10). Four-bank keyboard

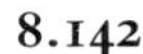

8.142

8.143

8.144

.142 RELIANCE VISIBLE (1915)
Pittsburgh Writing Machine Company, U.S.A. Also sold as PITTSBURGH, AMERICAN, BECKER BEACHLER, and RELIANCE-PREMIER. (Group 10). Four-bank

.143 REPORTERS SPECIAL (1914)
Rex Typewriter Company, U.S.A. Also sold as AUTOCRAT, HARRIS, REX DEMOUNTABLE, and REX VISIBLE. (Group 10). Three-bank keyboard

.144 RHEINMETALL STANDARD (1935) Serial No. 51,818

8.145

8.146

8.147

.145 ROBERTS NINETY (1919)
Roberts Typewriter Company of Stamford, U.S.A. (Group 10). Three-bank. Production discontinued in 1924

.146 ROFA (1921)
Produced in Germany and Holland. (Group 6) Down-strike from the Front. Three-bank semi-circular keyboard

.147 SALTER (1913) George Salter & Company, Scale Manufacturers, West Bromwich, England. The model illustrated was manufactured in 1913. Most of the models were Group 6 Down-strike from the Front. The model shown here has four rows of keys. Previous models had three rows of keys and were also known as the PERFECT, RAPIDE, and ROYAL EXPRESS

8.148

8.149

8.150

.148 SALTER 5 (1892) British

.149 SATURN (1899)
Stauder, Zurich, Switzerland. It was neither a typebar machine nor was there a type-wheel. Only about 300 were ever made

.150 SECOR (1905)
Secor Typewriting Company, U.S.A. (Group 10). Four-bank. Occupying the old Williams factory the Company made 7,000 machines before it failed

8.151

8.152

8.153

8.151 SENTA (1914)
Frister & Rossman, Germany. Also known as the PRESTO and BALKAN. (Group 10). Three-bank. Production ceased in 1930

8.152 SHOLES VISIBLE (1901)
The last production of Christopher Latham Sholes Typewriter Company, U.S.A., invented by Louis Sholes

8.153 SHOLES ZALMON (1911)
Sholes Typewriter Company, U.S.A. Also produced under the name of ZALSHO and ACME by the Lawrence Manufacturing Company, London, England. (Group 10) Front-strike. Three-bank. Zalmon G. Sholes died in 1917 and the activities of the Company ceased

8.154

8.155

8.156

8.154 SMITH (Emerson) (1907)
Produced by bankrupt Emerson Typewriter Company, U.S.A. Controlled by Harry A. Smith. (Group 6) Side-strike

8.155 SMITH VISIBLE (1909)
A machine produced by the Yetman Typewriter Transmitter Company, U.S.A. (Group 10). Four-bank. Under the direction of J. L. Smith it was placed on the market with little success

8.156 STEARNS (1905)
(Group 10) Front-strike. Four rows of keys. A limited number were produced in a period of approximately ten years

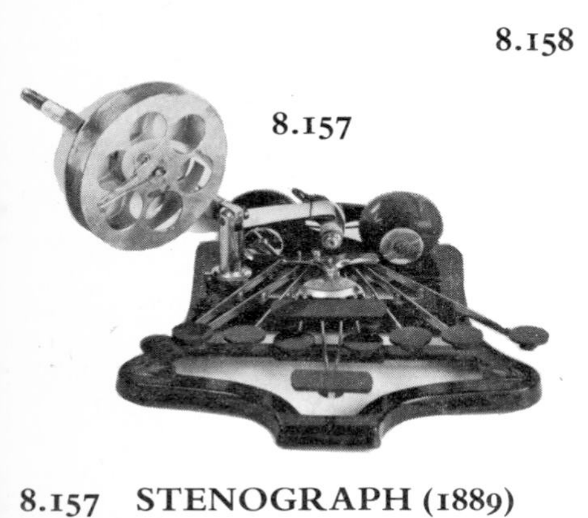

8.157

8.158

8.159

8.157 STENOGRAPH (1889)
United States Stenograph Company, U.S.A. A shorthand writing machine

8.158 STERLING (1910)
Sterling Typewriter Company, U.S.A. Partly Group 1 and partly Group 2. This machine used a shuttle. It was not successful and production ended in 1913

8.159 SUN (early experimental model) (1885)

8.160 8.161 8.162

160 SUN 2 (1901)
Sun Typewriter Company, U.S.A. It was also manufactured in Germany and sold as the STAR, NOVA, CARLEM, and LEFRALA. (Group 10). Three-bank. Nine different models were produced. It had limited success

161 TAURUS (1908)
A pocket typewriter. Very few were made or sold

162 TITANIA (1910)
Mix & Genest, Berlin, Germany. Since 1918 by Titania Typewriter Corporation, Germany. (Group 9) Front-strike with bearings side by side. Very similar to L. C. SMITH. Production ceased in 1925

8.163 8.164 8.165

.163 TRAVIS (1905)
Travis Typewriter Company, Philadelphia, U.S.A. (Group 2). It was not sold outside Philadelphia and the Company went bankrupt within six months

.164 VELOGRAPH (1886)
Invented by Adolph Eggis and manufactured in Geneva, Switzerland. It was the first Swiss typewriter. (Group 2). In some ways similar to the SIMPLEX toy typewriter later sold in America

.165 VELOGRAPH index typewriter (1887)
Presumably an improved model of the earlier VELOGRAPH. This machine has a circular disc and is a Group 2 index machine. It is very beautifully made and practical. Some models exist but very few apparently were ever sold

8.166

8.167

8.168

8.166 VICTOR (1927)
Victor Adding Machine Corporation, U.S.A. (Group 10). Four-bank. Production ceased after a few years

8.167 VICTOR 'A' (1894)
Tulton Manufacturing Company, U.S.A. (Group 5). Very few were made

8.168 VICTOR 'B' (1908)
Victor Typewriter Company, U.S.A. (Group 9). Bearings side by side. It had a poor reception and production ceased about 1917

8.169

8.170

8.171

8.169 VICTOR $10\frac{1}{2}$ (1919)
International Textbook Company, U.S.A. (Group 10) Front-strike. Four-bank. Presumably the $10\frac{1}{2}$ refers to the length of the rubber platen

8.170 VISIGRAPH (1910)
Visigraph Typewriter Company, U.S.A., and later C. Spiro Manufacturing Company, U.S.A. (Group 9). Four-bank keyboard with the line-spacer on the right-hand side. In 1919 it was produced by the Federal Typewriter Company which changed the name to FEDERAL. A few months later production ceased

8.171 WAVERLEY (1895)
A machine similar to the NORTH. (Group 6) Down-strike from Rear

8.172

8.173

8.175

8.172 WILLIAMS (1892)
Domestic Sewing Machine Company. U.S.A. (Group 7) Grasshopper Movement. This machine was the earlier model with three rows of keys

8.173 WILLIAMS 4 (1900)
Williams Typewriter Company, U.S.A. (Group 7) Grasshopper Movements. Four rows of keys. The printing was achieved by means of two ink pads, and the typebars moved in a peculiar manner very much like a Grasshopper's legs. The Company went bankrupt in 1909

8.174 WORLD 1 (1886)
Pope Manufacturing Company, U.S.A. (Group 5). Sold for approximately ten years and then the machine disappeared

8.174

8.176

8.1

8.175 WORLD 2 (1890)
Pope Manufacturing Company, U.S.A. (Group 5) Index System

8.176 WRIGHT SPEEDWRITER
No information as to its origin or date of production (Group 10) Front-strike. Four-bank

8.177 XCEL (Bennington's Syllabic)
Wesley Henry Bennington invented a syllabic writing machine and the second model was produced in April 1922. The Xcel is a similar improved machine writing syllables and short words in addition to the alphabet in upper- and lower-case. The Xcel Typewriter Corporation, New York, U.S.A. was formed by Mr. Bennington

8.178

8.179a and b

78 YETMAN (Transmitting typewriter) (1896)
Patented. (Group 9). Four-bank. It was designed to combine a telegraphic transmission instrument with a typewriter so that messages could be received and transcribed on the same machine

79a ZEROGRAPH (1895)
d b Produced by Mr. Leo Kamm of London. To the best of our knowledge only one of these exists. It was electric and produced in 1895. In his book on the subject George Carl Mares concludes with the following paragraph describing the machine which transmits telegraphic messages and foretells the future of Telex as follows: 'In view of the possibility of developments of this machine, therefore, there would seem to be no reason why a man sitting at his Zerograph in London, may not, in the future, be able to hold written converse with his correspondents in the furthermost parts of the globe, without the intervention of any actual physical connection.' This photograph was provided by Mr. Michael Adler of Italy, who claims he has the only machine in captivity. It shows that Mr. George Carl Mares was a prophet indeed!

b Carl Mares horse symbol

Conclusion

To gaze into the future is not really my province, but during the course of the production of this book I have continually been asked 'what is the future of the typewriter'?

The path of the prophet is a very thorny one indeed; but of two things I can be certain. Firstly, manufacturers cannot afford to stand still or be complacent.

As an example, in the 1920s the Underwood typewriter was so firmly established and selling so well that their salesmen were instructed to feign amazement if any prospective customer even suggested he might consider buying a different make of machine. They would challenge him to telephone any firm or office at once and ask them what make of typewriter they used; if the answer was not an Underwood, the salesmen would gracefully acknowledge defeat.

These tactics gave them at least a 50–50 chance as more than half of the typewriters then in use were Underwoods.

The Underwood Company continued with their well-established methods and were blind to the fact that Royal were progressing at an enormous rate and developing a first-class sales organization throughout the world. Whereas other manufacturers were producing machines with margin stops and a scale at the front, thus partially obscuring what was being typed, the Royal had a 'completely visible' typewriter with margin stops and a scale at the back—and later a segment shift—a design which made it very popular with typists who could see what they were typing.

By the end of the 1930s, Royal were selling 60 per cent of the machines sold in the world and Underwood were on the decline. Eventually through lack of foresight the Underwood Company was taken over by Olivetti and has now disappeared. With planning and adaptation to changing conditions they could still have been the best and foremost typewriter company in the world today.

Secondly, it is equally certain that great changes will come. No doubt less than a hundred years ago those who bought and sold horses and made carriages and harness for them felt perfectly secure in their occupation. After all, for thousands of years their business had undergone relatively few changes; little did they realize that in another thirty years or so, they were to be driven out of business by the motor car.

Today things move even faster and change much more rapidly; there has been more technological progress in the last hundred years than in the previous hundred thousand, and the rate of progress is increasing.

The typewriter is most unlikely to be an exception to this trend, and its basic conception and design, which has remained virtually the same since Wagner sold his machine to Underwood, must alter eventually. Perhaps we are on the verge of a revolutionary change such as has happened in recent years when mechanical calculators were superseded, almost overnight, by electronic ones.

It is possible that during the next decade typewriters will be manufactured in radio factories. The old notion of producing a typewriter to which you can speak and which will then type out letters automatically is a development in the very dim and distant future, not because it is a technical impossibility, but simply because those who dictate are themselves human beings, fallible and sometimes careless or inefficient, and often prefer to leave punctuation and other details to skilled and able secretaries.

Carl Mares in his book published in 1909 mentioned the Zerograph and in his final chapter, with prophetic wisdom, he foresaw the 'Telex', when the idea was hardly a conceivable possibility. With the accelerating rate of technological progress it is not easy to make accurate forecasts, but typewriters as we know them today, whether portable, manual, or electro-mechanical, seem to have a very limited life.

The Casio Computer Company Limited of Tokyo have developed an electronic typewriter which is the first single-element machine where no part of the element strikes the paper. Printing is achieved by the Casio Jet Printing System. With this machine it is possible to write 2,000 characters per minute. The type can be recorded on a magnetic tape or punch tape and the information can be re-printed automatically. The typewriter is completely silent.

Future development seems to lie in the direction of the electronic typewriter with electronic print-out, with no moving or mechanical parts whatsoever. The day may come when the letter typed will appear in front of the operator on a television screen in any selected type face, and by pressing a button, be transferred electronically to sensitized paper; alternatively, a process may yet be devised and developed which today might seem wildly impossible and completely impracticable, just as a few years ago none of us would ever have dreamed of sitting comfortably at home and seeing, in colour, men actually setting foot on the moon.

This book has dealt with facts as I believe they were or as I know them to be, and endless trouble has been taken to ascertain the accuracy of the statements made.

It is a comprehensive survey of typewriters to date. Should further information be required on any typewriter or matters relating thereto, the information will be provided, if available, by

British Typewriter Museum
137 Stewart Road
Bournemouth BH8 8PA
England

Please enclose, with your enquiry, a large envelope adequately stamped and self-addressed.

Bibliography

ADLER. *75 Jahre Adler* (Frankfurt: Adler-Werke).

ADLER, Michael H. *The Writing Machine* (London: George Allen & Unwin, 1973).

BLIVEN, Bruce, Jr. *The Wonderful Writing Machine* (Hartford: The Royal Typewriter Company Inc., 1954).

CURRENT, Richard Nelson. *The Typewriter and the Men Who Made It* (Ann Arbor: University of Illinois Press, 1954).

DUPONT, H. and CANET, L. F. *Les Machines à Ecrire* (Paris: Plume Sténogr., 1901).

GOULD, Rupert T. *The Story of the Typewriter* (London: Office Control and Management, 1948).

HELFER, Jürgen and MULLER, Philippe. *Konstruktionselemente der Schreibmaschine* (Aachen: Peter Basten, 1963).

HERRL, G. *The Carl P. Dietz Collection of Typewriters* (Milwaukee: The Board of Trustees, Milwaukee Public Museum, 1965).

JONES, C. Leroy 'Rocky'. *History of the Typewriter* (Springfield: Rocky's Technical Publications, 1956).

JONES, C. Leroy 'Rocky'. *Typewriters Unlimited—History of the Typewriter* (Springfield: Rocky's Technical Publications, 1956).

KRCAL, Richard and BASTEN, Peter. *Peter Mitterhofer and his Typewriter 1864 to 1964* (Aachen: Basten International, 1964).

LANG-KRUGER. *Handbuch des Maschinenschreibens* (Darmstadt: Winkler, 1936).

MARES, George Carl. *History of the Typewriter* (London: Gilbert Pitman, 1909).

MARTIN, Ernst. *Die Schreibmaschine und ihre Entwicklungsgeschichte* (Aachen: Basten International, 1949).

OLIVETTI, (C) E. C. *Olivetti, 1908–1958* (English version by Milton Gendel) (Ivrea: C. Olivetti & C., 1958).

RICHARDS, G. Tilgham. *The History and Development of Typewriters* (London: H.M.S.O., 1964).

ZELLARS, John A. *The Typewriter: a Short History on its 75th Anniversary 1873–1948* (New York: The Newcomen Society of England, American Branch, 1948).

The Story of the Typewriter 1873–1923 (New York: Herkimer County Historical Society, 1923).

The Typewriter: History and Encyclopaedia. Reprinted from *Typewriter Topics*, 1923, October (New York: Business Equipment Publishing Company).

Acknowledgment of Sources

We gratefully acknowledge information generously supplied by the following and, in particular, permission to reproduce in this book copyright photographs and tables.

COMPANIES

R. C. Allen
7.3*b*; Table 16

Brother International, Japan

Business Equipment Co., New York, U.S.A.
1.9; 1.13; 1.14*a*, *b*; 1.15; 1.18; 1.24; 1.28; 4.18

Ralph C. Coxhead Corporation (VariTyper)
3.21*a*; Table 5

Erika (Ideal) Dresden, E. Germany
Table 3

Facit (Halda) Office Equipment Co. Ltd., Rochester, Kent, and Ätvidaberg, Sweden
3.12; 3.13; 3.14*a–f*, *h*; 3.15*b–e*; 3.16*a*, *e*; Table 4

Godrej & Boyce Manufacturing Co. Ltd., Bombay, India

Grundig-Bürotechnik GmbH, Nürnberg
3.2*b*; 3.4*j*

GTA
3.2*c*; 3.3*c*, *f*; 3.4*b*, *d*, *f*; 5.18; 5.19; 5.20; Table 2

Hammond (VariTyper) U.S.A.

Hermes (Paillard S.A., Switzerland)
3.26; 3.27; Table 6

IBM United Kingdom Ltd., London
4.4*d*; Table 7

Imperial Typewriter Co. Ltd., Leicester
4.14; Table 8

Litton Industries Inc., U.S.A.
Table 14

Lucznik Zaklady, Metalowe, Warsaw, Poland

Messa (Siemag) Lisbon, Portugal
Table 9

Nippo Machine Co. Ltd., Yokohama, Japan
4.15

Nippon Remington Rand Kaisha Ltd., Tokyo, Japan
4.17*b*

Office and Electronic Machines, London
Ing. C. Olivetti & Company S.p.A., Italy
4.20*f*; 4.22*c*; Table 10

Olympia International, Germany
2.10–2.302; 2.304–2.329; 5.1; 5.2; 5.3; 5.4; Table 11

Optima Büromaschinenwerk, Erfurt, E. Germany
4.24

W. Rolf Pedersen A/S
2.331

Re-manufactured Typewriters Ltd., Angmering-on-Sea, Sussex
7.19*d*; Table 21

Remington Rand, London and U.S.A.
5.8; 5.9; Table 15

Royal Typewriter Co., Connecticut, U.S.A.
5.10; 5.11*a–c, e, f*; 5.12*a, b, d–j*; Table 12

Smith Corona (L. C. Smith, London, England and U.S.A.)
5.14; 5.15; 5.16; 5.17; Table 1; Table 13

Triumph Adler Vertrieb, West Germany
3.2*a*

Zeta, Zbrojovka Works, Brno, Czechoslovakia

MUSEUMS

Daimler-Benz Motor Museum, Stuttgart, Germany

Deutsches Museum, Munich, Germany
1.21

Forschungs- und Ausbildungstätte für Kurzschrift und Maschinenschreiben, Bayreuth, Germany

Leicester Museum, England
1.2

Institut National de Sténodactylographie Meysmans, Brussels, Belgium

Milwaukee Public Museum, Milwaukee, Wisconsin, U.S.A.
1.32; 1.37; 1.42

Schrift- en Schrijfmachinemuseum, Tilburg, Holland

Science Museum, South Kensington, London (Crown copyright)
1.1; 1.4; 1.6*a–c*; 1.8; 1.12; 1.16*a–c*; 1.17; 1.22*c*, 1.23; 1.43; 3.10*a*; 4.2*b*, 4.6*a, b*; 5.7*c, d*; 6.4*a*; 7.4; 7.14*a, b*; 7.21*a*; 7.22; 8.42; 8.49; 8.66; 8.68; 8.78; 8.83; 8.85; 8.111; 8.129; 8.130; 8.147; 8.161; 8.165; 8.171; 8.172; Table 23

Smithsonian Institution, Washington D.C., U.S.A.
1.19; 1.36

Technisches Museum, Vienna, Austria

Tekniska Museet, Stockholm, Sweden
1.46; 8.131

PHOTOGRAPHERS

J. Bauty, Lausanne, Switzerland
3.22*g*; 3.25*d*

Mitter Bedi, Bombay, India
3.18; 3.19

Gianni Berengo-Gardin, Milan, Italy
4.23

Arthur Coleman Ltd., Bournemouth
4.20*c*

Colore industriale di Alberto Fioravanti, Milan, Italy
4.20*e*

Fox Photos Ltd., London
3.3*h*

Goring Hill Studios, Goring-by-Sea, Sussex
7.19*c*

Atelier Rolf Günther, Dresden
3.11*e, h*

Gerhard Hickmann, Dresden
3.11*a–d*

Ken Hoskin of J. W. Kitchenham Ltd., Bournemouth
5.11*d*; 5.12*c*

Charles Jenkins, London
4.22*d*

N. K. Lali, Bombay, India
3.17*b*

Photolight-Studio, Milan, Italy
4.21*e*

Rooks Photo, Grand Rapids, Michigan, U.S.A.
7.3*a*

Heinz von Sterneck, Stockholm, Sweden
3.16*b, d*

Studio Granath, Caslon Press
1.22*a*

Svante Erixon Foto, Stockholm, Sweden
3.14*g*

Vasari Roma, Italy
8.179

Arne Wahlberg, Stockholm. Sweden
3.14*i*

Ward Hart, London
4.21*b*

Zakland Foto-Service, Poland
4.11*b*

PUBLISHERS

Austin Seven Instruction Book, 1922
5.6

Herkimer County Historical Society, New York, U.S.A.
1.47

The International Trade Paper, 1905, Nov.
1.26

Loving Magazine, I.P.C. Magazines, London
1.5

Gilbert Pitman, 85 Fleet St., London, E.C.
8.179*b*

Random House Inc., New York
6.11*a*

U.S. Patent Office Gazette
1.27; 1.38

Index

Addressograph-Multigraph Ltd., 113
Adler, 92–8
  merged with Triumph, 172
  production statistics, 96–7
Allen, R. C., 192–5
  production statistics, 194
Alphabets, 2
  keyboard, 43–4, 45–72
Arrington, George, 176

Barlock, 195–7
  production statistics, 198
Beach, Alfred E., 25
Bernadotte, Siggaard, 106
Bibliography, vii, 268
Bidet, Gustave, 9
Blickensderfer, 177, 199–200
Blind typewriters, 12–13, 16, 17, 179, 183
Bluebird production statistics, 212–13
Braille, Louis, 12, 183
British Typewriter Museum, vii, 267
Brother, 89, 98–100
Brown, Alexander, 161, 163
Brunsviga, 151
Burroughs, 200–1
Burstein, Max, 82
Burt, William Austin, 4–7
Byron, 197
  production statistics, 198

Cahill, Thaddius, 176
Caligraph, 165
Carbon paper, 9, 14
  for copy purposes, 84
  used instead of ribbons, 83–4
Carbon ribbons, 84–5, 112
  attachments, 85–6
Casio Computer Co. Ltd., 267
Caxton, William, 3
Chinese characters, 138
Chinese typewriters, 189–90
Cipher machines, 185–6
Clough, Jefferson M., 32
Colby, C. C., 183
Composing machines, 112–13
Computer design, 159
Computer output typewriters, 127
Conqueror, 89, 210
Consul, 175
Conti, Pietro, 8
Continental, 201–2
  production statistics, 202
Cooper, John, 25
Corona
  use during World War I, 37
  see also Smith-Corona
Coxhead, Ralph C., 111
Cryptography, 185
Cuneiform script, 2

Daugherty, James Denny, 27
Deaf and dumb people, 22
De Mey, F. A., 25
Densmore
  merger with Smith-Premier, 165
  Wagner's contribution, 26
Densmore, James, 28, 29–32
De Sondalo, Baillet, 10
Devincenzi, Giuseppe, 176
Driesslein, C. L., 176
Dujardin, A., 9, 185
Durability of typewriters, 87–8
Dvorak, August, 42
Dynacord production statistics, 212–13

Edison, Thomas A., 30–1, 176
Electric typewriters, 122–3, 176–8, 180–3
  annual sales, 182
  early developments, 176–7
Electromatic Typewriters Inc., 122
Electronic typewriters, 267
Embossed characters, 12
Ennis, G. H., 177
Erika, 100–2, 104, 147

Eumig, 114
Everest, 202–3
  production statistics, 204

Facit-Halda, 89, 104–8
  production statistics, 107
Fischer, Oscar, 180
Flat surface typewriters, 10–11
Forward-stroke machines, 27
Foucault, Pierre, 13

Gabrielson, Carl, 165
Glidden, Carlos, 25, 28
Godrej, 108, 109
'Golf ball' typewriters, 39, 125
  introduced by IBM, 127
Grundig, Max, 98, 72
Guillemot, Adolphe Charles, 16–17
Gutenberg Bible, 3

Halda production statistics, 107
Halda Fickursfabrik, 104
Hall, Thomas, 25
Hammarlund, Henning, 104–5
Hammond, 108, 110–11
Hammond, James B., 24–5, 108, 110–11
Handicapped people, *see* Blind typewriters; Deaf and dumb people
Hansen, Rasmus Hans Johan Malling, 21–3, 176
Hermes, 89, 113–21
  chronology, 114
  production statistics, 118
  type faces, 120–1
Hess, E. B., 157
Hood, Peter, 15
House, George, 25
Hughes, Mr., 12

IBM Corporation, 89, 122–7
  computer production, 124
  'golf ball,' 39, 125, 127
  production statistics, 123
  Selectric machines, 127, 181–2
Ideal, 100–2, 104
Imperial Typewriter Co., 127–9, 161
  begins production, 128
  production statistics, 130–1
  Queen's Award for Industry, 129
Interchangeable type, 25

Japanese characters, 138
Jenne, William K., 32, 152
Jones, John M., 25
Juventa, 202–3

Kempelen, Wolfgang, 7–8
Key lever linkage with typebar, 26
Keyboards
  discussions on standard layout, 41
  early arrangements, 39–41
  modern layouts (alphabetically by language), 45–72
  nineteenth-century layouts, 43–4
  plastic, 88
  problems of non-standardization, 165
  'QWERTY' arrangement, 41, 43
  simplified, 42
  slope, 179, 183
  special symbols, 73–5
Keys
  design, 88
  typamatic, 127
Kidder, Wellington Parker, 28, 183
Kleyer, Heinrich, 92
Kovo, 90

Legal acceptance, 82–3
Letter use frequency, 41–2
Letters of the alphabet, 90
Line spacing, 21–3, 112
Litton Industries Corporation, 89, 129
Livermore, Benjamin, 25
Lucznik, 135

McGurrin, Frank E., 40
Machine-readable numerals, 91
Manufacturers
  gone out of production, 89
  producing currently, 89–90
  total output, 88
Marchant Calculators, 169
Mercedes, 203, 205
Messa, 90, 136, 139
Mill, Henry, 3
Mitterhofer, Peter, 18–21
Moya, Hidalgo, 127–8
Museum, British Typewriter, vii, 267
Music typewriters, 11, 181, 186–7

Neipperg, Leopold von, 7, 15
Nippo Machine Co. Ltd., 137–8, 139
Nippon Typewriter Co. Ltd., 138, 139–40
Noiseless, 208–9
Noiseless typewriters, 179, 183–5
Norica, 169
Numerals, 90–1
  machine readable, 91

Oliver, 205–8
  production statistics, 206

Oliver, Thomas, 205
Olivetti, Camillo, 140
Olivetti Company, 140–7
  production statistics, 142–3
  take over Underwood, 218
Olivetti/Underwood Corporation, 89, 218
Olympia, 89, 147, 149–52
  production statistics, 154–5
Optima, 147

Paillard, Moïse, 113
Papyrus, 1–2
Parchment, 2
Peeler, Abner, 25
Pellaton, George, 181
Pens, 2
Perrot, Louis Jérôme, 9
Petty, William, 3
Pitch, 78–82
Plastic keyboards, 88
Plastic ribbons, 85
Pneumatic typewriters, 180, 186
Portable typewriters, 27, 199
Pratt, John, 23–5
Printing, 2–3
Production figures, 86–7
Progin, Xavier, 9, 10–11
Proportional spacing, 124, 143
Pterotype, 23, 24

Quills, 2
'QWERTY' keyboards, 41, 43

Ranson, James, 3
Ravizza, Giuseppe, 9, 14–15
Rebuilt typewriters, 188–90
Remington, 152–3, 155–6, 157–8
  first electric model shown, 180
  merger with Smith-Premier, 165
  started production, 88
  Wagner's contribution, 26
Remington, Philo, 31, 32
Remington Rand Ltd., 89
Ribbon spools, 84
Ribbons, 83–6
  carbon, 84–5, 112
  cushion substitute, 27
  early travelling inked, 14
  plastic, 85
  widths, 84
Rose, Frank, 166
Rose Typewriter Co., 166–7
Royal, 157–61, 164–5, 166
  production statistics, 160–1

S.C.M., *see* Smith-Corona
Sales figures, 86–7
Schlüns, Karl, 203
Schueler, Franz, 203
Seidel and Naumann, 100–2, 104
Selectric, 127, 181–2
Serial numbers, 221
Sheldon, John, 5
Shift keys, 11, 15
Sholes, Christopher Latham, 14, 24, 25, 28–32, 152
Shorthand, 34
Shorthand writing machines, 182, 187
Siegener Maschinenbau A.G., 136
Siemag/Messa, 90, 136, 139
Siemag production statistics, 137
Silenta, 90, 201–2
Smathers, James F., 122
Smith, H. W., 167, 169
Smith, Lyman and Wilbert, 161, 163, 165, 167
Smith-Corona, 89, 161–3, 165–9
  production statistics, 162–3
  see also Corona
Smith-Corona Marchant Inc., 169
Smith-Premier, 163, 208–9
Social aspects of the typewriter, 13, 32–8
Soule, S. W., 25, 28
Spacing
  proportional, 124, 143
  variable, 113
Speed typing, 39–41
Spools, *see* Ribbon spools
Standard Typewriter Co., 166–7
Stoewer, 89, 209–10
Stone Age drawings, 1
Sugimoto, Kyota, 138
Swift, 210
Symbols, special, 73–5

Tabulators, 113
  electric, 127
Taub, Louis, 40–1
Telegraphic type printers, 15
Thompson, Russell, G., 122
Ticker-tape machines, 31
Torpedo, 210–13
  production statistics, 212–13
Touch typing, 40
Toy typewriters, 14, 187–8
Triumph, 169–73
  merged with Adler, 172
  production statistics, 170
Turri, Pellegrino, 8
Typamatic keys, 127

Type
 body measurements, 83
 curvature, 82
 main producers, 78
Type faces, 78–81
 Hermes typewriters, 120–1
 interchangeable, 25, 111–13, 125
 upper and lower cases first used, 152
Typebars
 first recorded use, 10
 interchangeable, 127
 linkage with key lever, 26
Typewriters
 alphabetical list by model name, 226–45, *and illustrated*, 246–65
 categories, 222–5
 durability, 87–8
 early American, 4–7
 early European, 7–10
 electric, *see* Electric typewriters
 electronic, 267
 music, 11, 181, 186–7
 pneumatic, 180, 186
 rebuilt, 188–90
 shorthand writing, 182, 187
 social stigma, 13
 toy, 14, 187–8
 use by the professions, 35–7
 use during the World Wars, 37–8
Typewriting
 expert analysis, 82
 first use of word, 24
Typists, 34–6
 early training, 35
Typographer, 4–7

Uhlrig, R. W., 192
Underwood, 214–18
 designed by Wagner, 26, 214
 production statistics, 217–18
 taken over by Olivetti, 143, 218
Union Typewriter Co., 165

VariTyper, 111–13
 chronology, 112
 first produced, 180
Vellum, 2
Visible Writing Machine Co., 159

Wagner, Franz Xavier, 26, 214
Watson, Thomas J., 122
Wheatstone, Charles, 15
Women as typists, 34–5
Woodstock, 192–5
 operations during World War II, 37
 production statistics, 194
Writing Ball, 21–3

Yeremias, Arnold, 176
Yost, George Washington Newton, 27, 31, 219
 merger with Smith-Premier, 165
 Wagner's contribution, 26
Young, James, 3

Zbrojovka Works, 171, 173–5
Zeta, 173, 175